Observatories of the Southwest

Observatories of the Southwest

A Guide for Curious Skywatchers

Douglas Isbell and Stephen E. Strom

The University of Arizona Press Tucson

The University of Arizona Press

www.uapress.arizona.edu

Library of Congress Cataloging-in-Publication Data
Isbell, Douglas, 1964–
Observatories of the Southwest : a guide for curious skywatchers / Douglas Isbell and Stephen E. Strom.
p. cm.
Includes index.
ISBN 978-0-8165-2641-3 (pbk. : alk. paper)
1. Astronomical observatories—Southwestern States—History. 2. Astronomical observatories—Southwestern States—Guidebooks. 3. Southwestern States—Guidebooks. I. Strom, Stephen E. (Stephen Eric), 1942– II. Title.
QB82.U6I83 2009
522'.1973—dc22

2009024799

Manufactured in the United States of America on acid-free, archival-quality paper containing a minimum of 30% postconsumer waste and processed chlorine free.

Contents

Illustrations

Fred Lawrence Whipple Observatory

National Radio Astronomy Observatory Very Large Array

The Observatories of Sacramento Peak

McDonald Observatory

Mount Graham International Observatory

Preface

Stephen E. Strom

My passion for astronomy began when I was seven, inspired by a well-written entry in a children's encyclopedia, *The Book of Knowledge.* I soon began haunting a branch of the New York Public Library located a few blocks from the apartment house in which I grew up, searching first for books that would provide a guide to objects in the night sky, and later, for popular summaries of ongoing research by astronomers and physicists such as Fred Hoyle, James Jeans, and George Gamow. My reading, coupled with trips to the Hayden Planetarium too numerous to count, further stoked my interest in astronomy.

Among the most influential books in my early reading was David Woodbury's *The Glass Giant of Palomar,* which vividly described the twenty-year effort to design, fund, and construct the 200-inch telescope on Palomar Mountain, north of San Diego. The imposing power of the 200-inch to probe the farthest reaches of the Universe captured my imagination forever. I vowed at the age of ten or eleven that I would someday make use of the 200-inch—in retrospect an arrogant thought for someone so young but one that inspired me for the better part of the next dozen years until that childhood dream was realized.

Somehow, my parents recognized that my early interest was far more than casual and, following a multitude of entreaties, I was given my first telescope: a refractor with a lens one inch in diameter, about as powerful as the first instrument Galileo used in the seventeenth century to carry out his pioneering and courageous explorations of the heavens. That first telescope, along with various guidebooks and star atlases, provided me with a tool that occupied me for countless evenings and early-morning hours, as I searched the skies over New York from the roof of a "Bronx box" for double stars, star clusters, and the few galaxies that could overcome the bright glow of city skies.

By age thirteen, I had long since outgrown my tiny refractor and began to research how to build larger telescopes. Following a year of grinding and polishing a six-inch mirror at the optical shop of the Hayden Planetarium, I completed the first of several telescopes that I constructed before leaving for college in 1958. At about the time I had achieved "first light" with my six-inch, I had the enormous good fortune of having an opportunity to visit the Mount Wilson Observatory near Los Angeles and, furthermore, to talk with its director, Horace Babcock. (Life was simpler in 1956, so much so that an observatory director could afford to spend part of an afternoon chatting with a callow fourteen-year-old about magnetic field measurements of the Sun and the workings of an automated photoelectric guider to keep telescopes pointed at their targets—and to lead a tour of the legendary 100-inch and 60-inch telescopes.)

This random combination of early introductions to astronomy—involving natural curiosity, readily available first-rate written references, supportive parents, and the great fortune of visiting a *real* observatory at an impressionable age—resulted in a lifelong commitment to a career as a research astronomer and teacher. This career has not only fulfilled my childhood dreams, but instilled something far deeper: insights into the complex workings of a Universe whose beauty and mystery continue to both fascinate and awe.

Over my professional lifetime, I have traveled the world to use optical and radio telescopes and have had the great fortune to work with hundreds of memorable colleagues, many of whose passions in astronomy were aroused and nurtured in much the same way as mine.

The motivation for this book was to provide a guide for today's young women and men whose curiosity about the world about them has led them to consider a career in astronomy or other sciences, and for "amateur" astronomers (in the best sense of that word) who wish to pursue their interests further. In the end, we hope it will appeal to anyone who has a fascination with astronomy and space science and a curiosity about humanity's place in the vast Universe and our ultimate fate.

This book is the result of a close collaboration between me and Doug Isbell, a talented science writer and public affairs professional who has dedicated his life to bringing the excitement of discoveries in astronomy and space science to the public. Doug has worked with his counterparts at other astronomical institutions in the southwestern United States to offer readers an overview of the rich history of eight observatories located in Texas, New Mexico, Arizona, and California and the powerful instruments that sit atop many of their majestic mountaintops.

Through talks with active research scientists, Doug also captures the "flavor" of observing at each of these sites—much as Horace Babcock provided me through his generous sharing of an afternoon more than fifty years ago—and offers a glimpse into the complex personalities of the women and men who have led lives of intertwined personal and professional discovery. I have tried to add selected summaries of the exciting research carried out at each of these institutions and how that science fits into the larger themes of modern astronomical investigations.

Together, we aim to provide readers with the information and inspiration needed to spark a visit to these wonderful places and gain a richer experience while there. It is our hope that all who read this book will take the time to visit at least one observatory, and gaze with wonder upon the best that women and men can do in response to our shared yearning to connect with the true history of our solar system and the awesome Universe beyond.

Abbreviations

AGN	active galactic nuclei
ALMA	Atacama Large Millimeter Array
APOLLO	Apache Point Observatory Lunar Laser Ranger Operation
ARC	Astrophysical Research Consortium
ATST	Advanced Technology Solar Telescope
AURA	Association of Universities for Research in Astronomy
CCD	charge-coupled device
CfA	Harvard-Smithsonian Center for Astrophysics
CTIO	Cerro Tololo Inter-American Observatory
DCT	Discovery Channel Telescope
EVLA	Expanded Very Large Array
HET	Hobby-Eberly Telescope
IOTA	Infrared-Optical Telescope Array
KBO	Kuiper Belt object
LBT	Large Binocular Telescope
LOR	Large Optical Reflector (Telescope)
MMT	Multiple Mirror Telescope
MONET	MOnitoring NEtwork of Telescopes
NASA	National Aeronautics and Space Administration
NOAO	National Optical Astronomy Observatory
NPOI	Naval Prototype Optical Interferometer
NRAO	National Radio Astronomy Observatory

NSF	National Science Foundation
NSO	National Solar Observatory
QSO	quasi-stellar object
RET	Research Experiences for Teachers
REU	Research Experiences for Undergraduates
SAO	Smithsonian Astrophysical Observatory
SARA	Southeastern Association for Research in Astronomy
SMT	SubMillimeter Telescope
TMT	Thirty Meter Telescope
VATT	Vatican Advanced Technology Telescope
VERITAS	Very Energetic Radiation Imaging Telescope Array System
VLA	Very Large Array
VLBA	Very Long Baseline Array
WIYN	Telescope operated by the University of Wisconsin, Indiana University, Yale University, and NOAO

Observatories of the Southwest

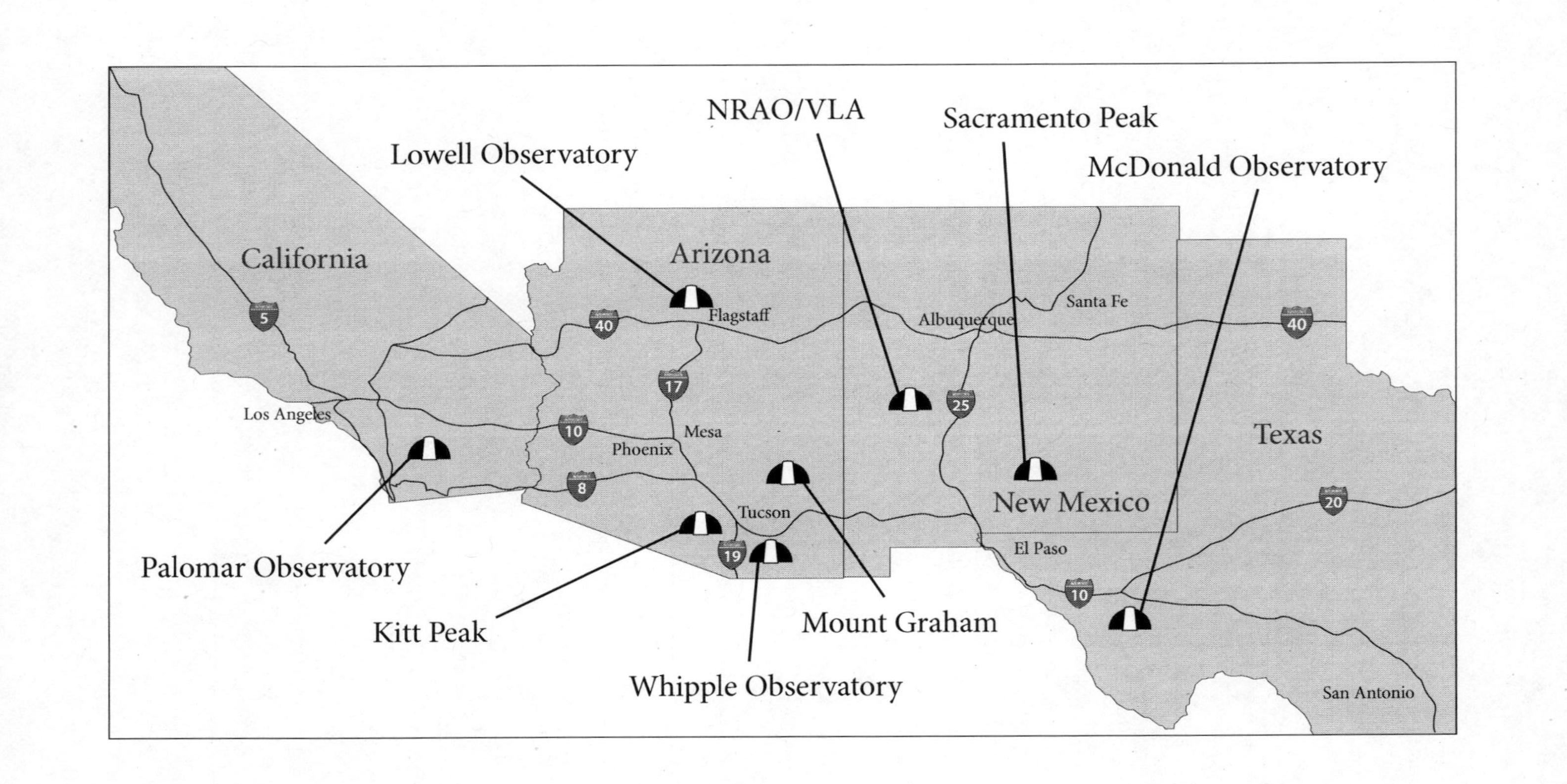
NRAO/VLA
Sacramento Peak
Lowell Observatory
McDonald Observatory
California
Arizona
5
40
Flagstaff
Santa Fe
Albuquerque
40
17
25
Los Angeles
10
Mesa
Phoenix
Texas
8
New Mexico
Tucson
20
El Paso
19
10
Palomar Observatory
Kitt Peak
Mount Graham
Whipple Observatory
San Antonio

Introduction

Astronomical observatories are special places. They share the common goal of seeking deeper understanding of the Universe that surrounds our tiny home planet, but each has its own unique array of equipment and historical lineage. Over time, observatories have morphed from isolated outposts populated by behemoths of steel and glass to high-tech complexes filled with sophisticated imaging and computer-based analysis tools, linked continuously to a worldwide network of telescopes and rich digital archives.

Telescope mirrors have grown lighter and dramatically thinner. (And note that in this book we define the size of a telescope's primary mirror in both English and metric units on first reference, and then defer to the most common modern usage.) Observing tools have changed profoundly. Fifty years ago, the main tools of the astronomer were large photographic plates, which recorded images of the sky gathered with amazing patience by hardy individuals riding in steel cages attached to the telescope superstructure—often in frigid weather conditions. Today, digital cameras that produce files of billions and billions of bits (gathered and analyzed instantly in warm control rooms) have replaced photographic plates, while ever more specialized instruments exploit advances in detector and computer technology to dissect cosmic light in ever more sophisticated ways. These modern digital images and spectral (highly detailed color) maps enable analysis of the Universe beyond Earth with a clarity and sensitivity that would surprise, and perhaps even shock, earlier generations. These previous generations of astronomers—largely men—might also be surprised by the gender revolution in astronomy, which has grown to embrace the talents and imaginations of women and in the process has transformed astronomy into one of the most diverse fields of the physical sciences.

With these new and powerful tools, astronomers are now able to probe the Universe back to the dawn of time, peer into the birthplaces of stars, and study strings of galaxies stretching across hundreds of thousands of light-years.

As a result, astronomers today are able to confront some of the most profound questions regarding our origin as humans and our ultimate fate as a lonely planet in one tiny part of a vast Universe. Throughout its history as a science, astronomy has consistently revealed and challenged the most fundamental conclusions about the basic properties of the Universe and the place of humans within it. Indeed, surprises, new discoveries, and paradigm shifts have accelerated as the power of our tools of exploration has increased.

One recent example is a discovery that has shaken modern cosmology to its core.

In 1998, two competing teams simultaneously announced similar supersensitive and double-checked measurements of the brightness of known types of medium-sized exploding stars known as supernovae, housed in distant galaxies. Each team was initially surprised to find that the supernovae were slightly farther away than the best calculations said they should be. After reviewing their data, and exploring a range of conventional possibilities to explain this discrepancy, they were forced to posit the existence of a mysterious force, dubbed "dark energy," that is pushing all of the galaxies and other matter in the Universe apart from one another, at an accelerating speed. What this force might be is completely unknown. The hunt to solve the "mystery" of dark energy drives curious scientists to propose and make new observations that stretch the state of the art and hold the potential of radically changing the way we understand our physical world.

Using some of the same telescopes (along with many others) and some clever data analysis tools, astronomers have recently discovered planetary bodies in our own solar system the size of Pluto and larger; hundreds of nearby stars surrounded by planets; and huge numbers of strange, dimly glowing, nearly starlike hulks known as brown dwarfs.

One group has already staked a claim to taking an image of the first realistic cousin to the gas giant planet Jupiter. Pictures of Earth-like planets may populate the Web pages of tomorrow in fewer than twenty years.

Observatories have also evolved dramatically in recent decades, from entities created by wealthy and well-connected individuals and universities to an egalitarian mixture of public, private, and international facilities sprinkled around the globe on mountains, sand, and ice. Astronomers have deployed telescopes and detectors on Earth, aboard balloons, in Earth orbit, and in distant space that study the Universe not only in optical light—where our eyes operate—but across nearly the entire radiation spectrum, from super-high-energy gamma rays with minuscule wavelength to radio waves longer than a football field.

The southwestern United States is unique in this loose federation of like-minded research outposts, in both the quantity and the diversity of its astronomical observatories. Urbanization of the surrounding communities has encroached some upon their dark skies, but the best conditions remain world class, and the natural beauty of the land that surrounds them remains awe inspiring.

This guide is aimed at providing an introduction to these observatories for those eager to see the tools of the astronomer up close, learn what modern telescopes and their sophisticated instruments are doing today, and understand a bit about the role each of these observatories has played in advancing human understanding of the cosmos so far in just the past dozen decades. As such, it seeks to aid both the interested student and the curious adult in planning a trip to one or all of these observatories—a trip that we hope may be the beginning of a longer journey or quest.

The 200-inch telescope in the moonlight.

Palomar Observatory

North San Diego County, California

Home of the Giant

The massive 200-inch (5.1-meter) Hale Telescope on Palomar Mountain in southern California is an icon of mid-twentieth-century scientific and engineering prowess. The "200-inch," as astronomers are prone to call it in shorthand, was born in an age when the ability to dream up fantastic new machines—and to be able to sell them to the powers that be—dwarfed any serious concerns that it was too difficult or costly to create them.

The mammoth effort to design and build the Hale Telescope tapped into the zeitgeist of the 1920s and 1930s. The nation-spanning pursuit was stoked by the very public nature of the quest of leading scientific figures to understand our place in the cosmos. Edwin Hubble's photographic plates from the 100-inch (2.5-meter) Hooker Telescope on Mount Wilson near Los Angeles showed fuzzy "island universes" (galaxies similar to the Milky Way) apparent by the dozens. But what was the true scale of the Universe beyond these relatively nearby galaxies? Just how vast was it? Were galaxies far, far away really receding from Earth with a velocity directly proportional to distance, as implied by

Web site
www.palomar-observatory.org

Phone
(760) 742-2119
(Public Information Recording)

Address
35899 Canfield Road
Palomar Mountain, CA 92060
Directions & maps posted at:
www.astro.caltech.edu/palomar/driving.html

Hubble's straight-line graph? To answer these fundamental questions, a group of well-connected astronomers with vivid personalities banded together and sought the funds to build a machine of a size matched to the challenges.

The drive to build the 200-inch telescope was fed by the growing public regard for scientific progress, engendered by unlikely celebrities such as Hubble and Albert Einstein. It also drew heavily upon the new vision of transformative social engineering promised by the large funding bequests being dispersed regularly by emerging mega-foundations like the Carnegie Institution. The fact that it would require enough resources to build a battleship-sized blend of metal and glass capable of moving and pointing with better precision than a wristwatch was hardly a deterrent to its mastermind, the visionary midwestern astronomer George Ellery Hale.

A wealthy native of Chicago, Hale was a solar astronomer by training (at the Massachusetts Institute of Technology [MIT] and Harvard Observatory, with a stopover in Berlin). As his first major achievement, Hale founded several venerable journals of astronomical research that continue to be published today (and later was instrumental in the formation of the National Research Council). Then, largely because he *could* do it, he used money from private benefactors whom he met and wooed in Chicago's high society to build Yerkes Observatory, operated by the University of Chicago, in southern Wisconsin.

Yerkes was dedicated in October 1897 (after major repairs owing to the collapse of the original floor five months before) and, for a brief time, it was the most powerful nighttime astronomical observatory in the world. When the Yerkes telescope was completed, Hale was just twenty-nine years old. The mounting and tube for the 40-inch (1-meter) telescope at Yerkes were a marvel of the legendary 1893 Columbian Exposition in Chicago, and the 40-inch remains the world's largest refracting telescope (bending the incoming starlight through a giant glass lens rather than reflecting it off large curved mirrors).

But the plaster was barely dry on the buildings at Yerkes before Hale's passion for larger and larger telescopes brought him his first success with the Carnegie Foundation—grant funding for a 60-inch telescope that became the founding element of legendary Mount Wilson

Observatory, located in the mountains above Pasadena, California, outside Los Angeles.

A largely loveless marriage and several improperly diagnosed medical maladies (probably rooted in an inner-ear disorder) caused major physical strains on Hale. He suffered from tremendous headaches and dizziness between manic bouts of technical work mixed with endless cajoling of supporters and potential benefactors. Hale died in 1938 at age sixty-nine, a decade before the 200-inch was complete, but his bullish passion had given the project more than enough momentum to survive his poignant passing.

Design and construction of the giant Palomar telescope began in 1929, continued throughout the Great Depression, and survived a four-year hiatus during World War II. By the time this "perfect machine" first opened its shutter for science on November 12, 1949, the 200-inch telescope had challenged nearly every engineering process of the era. The grinding and polishing of the mirror alone took eleven years, after the second attempt by Corning Glass Works to pour the primary mirror using a new glass called Pyrex was finally deemed acceptable. This followed a Herculean failed effort by industrial giant General Electric to fabricate the huge mirror blank from a molten sleet storm of fused quartz, heated to 4,000 degrees and sprayed into a mold by a fiendishly complex bank of nozzles within a massive furnace.

The eventual success of the 200-inch telescope project became a point of national pride. The excitement over the giant telescope project was amplified by the emerging power of the national media, both print and radio, and by the widening name recognition of the Rockefeller Foundation, which both enabled and endorsed the project with an initial $6 million grant (about $70 million in early-twenty-first-century dollars). National fervor over the project reached its peak in 1936 with the cross-country train trip of the unpolished 200-inch mirror blank over a carefully planned patchwork of the country's still-unwieldy regional networks of commercial railway lines, bridges, and tunnels. The specially outfitted train car with the tilted mirror stored inside drew large, enthusiastic crowds on the order of the latest presidential candidate or singing idol.

Many design elements of the 200-inch telescope—such as its horseshoe structure, forced-oil bearings, and Surrier truss—were subsequently used in a variety of large telescopes, including the 158-inch (4-meter) Mayall Telescope at Kitt Peak. These innovations emerged over two decades from the brains, hands, and slide rules of a variety of project engineers and managers. The tedious work of polishing and optically checking the mirror, combined with the often heavy hand of shop director Marcus Brown, made working in the Caltech optical shop a battle of high-intensity boredom. (Although even a casual visit to the optical shop to witness the grandeur of the emerging mirror was enough to give inspiration to future leaders and entrepreneurs, such as young Aden Meinel, who went on to influence both Kitt Peak National Observatory and Whipple Observatory.)

The job of erecting the observatory's buildings and the 200-inch dome on ruggedly remote Palomar Mountain, led by retired World War I army colonel M. L. Brett, fell largely to a patchwork crew of grateful Depression-era laborers and Caltech grad students, who dug and poured cement by hand and wheelbarrow when necessary.

No single person beyond Hale was more influential in the creation of the 200-inch than Russell Porter, an amateur astronomer and illustrator from Springfield, Vermont. Among other adventures, Porter was a veteran of numerous Arctic expeditions, and he founded the famous Stellafane star party held each summer in his hometown—an event still attended by the most passionate amateur astronomers. Porter had impressed Hale during his duties as a scientific journal editor with imaginatively detailed drawings of an instrument for splitting starlight into its component colors (known as a spectrograph).

Porter soon found himself working as a human mechanical-drawing machine on the 200-inch project, churning out hundreds of pencil-and-paper renderings of the giant telescope and its instruments. These hand drawings, which continue to be admired and sold as art posters today, simultaneously served as amazing visualizations of yet-to-be-built hardware and as functional engineering drawings that were so eerily accurate that they were used to construct and assemble the major systems of the gargantuan telescope. Particularly in cutaway views that show both the placid exterior and the complex inner workings of the telescope, Porter's exquisite drawings left an indelible mark on the popular image of what a telescope should look like.

The legendary Edwin Hubble never monopolized the 200-inch telescope nearly as much as he had expected, though he did take one of the earliest "deep" images of faint galaxies, duplicating in a mere six-minute exposure all of the detail that his treasured 100-inch Hooker Telescope on Mount Wilson had ever recorded in the region. Hubble was lastingly bitter that he lost out to Ira Bowen, a laboratory astrophysicist at Caltech, to serve as the first director of Palomar Observatory (Bowen was selected by legendary mid-twentieth-century science policy icon Vannevar Bush). Moreover, there turned out to be far too many interesting ideas and clever astronomers who competed for the Hale Telescope's precious observing time to allow Hubble all of the access he had dreamed of exploiting.

The first enduring landmark discovery with the 200-inch was published in 1956 by astronomer Walter Baade, who finally had the spatial resolution that he craved to discern and measure individual stars in the bright core of the nearby spiral galaxy Andromeda (also known as Messier 31, or M31, in the enduring eighteenth-century star catalog recorded by pioneering French sky mapper Charles Messier). This led him to deduce that there were two distinct populations of the special variable stars, known as Cepheids, that Hubble had been using to calibrate his galactic distance scale, rather than just one; when the proper adjustments were made to Hubble's calibrations, Andromeda was suddenly understood to be twice as far away as had been commonly believed (modern calculations of 2.5 million light-years between the Milky Way and Andromeda are surprisingly close to Baade's original estimate of the distance).

Ten years later, on March 11, 1966, Maarten Schmidt appeared on the cover of *Time* magazine for observations of an object known as 3C 273. This object appeared as a starlike point in the sky in visible-light images, but spectrographic observations with the 200-inch showed it to exhibit the signatures that we now understand to be characteristic of those expected from the staggeringly large energetic output of a black hole at the center of a galaxy as it actively pulls in nearby gas. Today such quasars (or active galactic nuclei, AGN) are known to exist by the hundreds, yet they remain enigmatic entities, the object of much study. 3C 273 is located more than 2.4 billion light-years from Earth, but it

The 200-inch Hale Telescope with adative-optics laser firing.

is so bright that it can be spotted as a faint starlike point in an 8-inch (0.2-meter) amateur telescope. It and its more distant brethren provide luminous beacons that permit astronomers to probe the most remote parts of the observable universe.

Today, the Hale 200-inch is outfitted with the cutting-edge technology of adaptive optics, using a laser beam to create a point of light high in the atmosphere that twinkles in a nearly identical manner to the incoming starlight. In most adaptive-optics systems, a flexible mirror located along the light path between the primary mirror and the detector has its surface shape adjusted constantly by small pistons at a rate of hundreds of times per second. These adjustments provide for real-time compensation for the distorting effects of Earth's atmosphere on the shape of the incoming "wavefront" of light from a cosmic source.

Time lapse view of the 48-inch Samuel Oschin Telescope showing star trails around the north celestial pole.

The 200-inch would probably not have achieved anything near its potential for discovery without its much smaller companion at Palomar, the 48-inch (1.2-meter) telescope. The 48-inch (named the Oschin Telescope today for modern benefactor Samuel Oschin) was slipped into the Rockefeller Foundation funding envelope of the larger telescope almost as an afterthought, but it was a vitally important addition, given that the massive telescope would need much better wide-field photographs of the night sky—that is, deep finding maps—than had ever been made. Surveys of the sky with the Oschin continue to provide astronomers with lists of target objects for more detailed study with the 200-inch and other large telescopes.

The 48-inch telescope represents one of the earliest realizations of a Schmidt telescope, named for the German scientist who first had the idea of adding a thin corrector lens over a simple spherically shaped primary mirror to produce fine images over a very wide field of view. The primary mirror of the Palomar 48-inch telescope is actually 72 inches wide (and was much easier to grind and polish than the more complex parabola required by more traditional reflectors), but the telescope's name is defined by the size of its clear 48-inch glass corrector lens.

The 48-inch was designed and built with such precision that its very first glass plate, containing an image of the nearby spiral galaxy Andromeda (M31), was of such high quality that it was published as part of Edwin Hubble's pioneering atlas of galaxies, the first complete bible of the multiplicity of galaxy "types" (spiral, elliptical, irregular) that populate the Universe. The telescope will be forever cited in history books for producing the National Geographic Palomar Sky Survey, a vast imaging database of the northern sky compiled from photographic plates made from 1949 to 1958 and repeated from 1985 to 2000.

The data from this unparalleled survey, now digitized, have been the source of 50 million newly discovered galaxies and a billion newly known stars. The Palomar Sky Survey remains so enduring that it forms the basis of Sky at Google Earth and is a basic layer of Microsoft's World-Wide Telescope software, making hundreds of nights of scientific toil and dedication available at the click of a computer mouse.

The most recent claim to fame for the 48-inch is its use by Michael Brown of Caltech and collaborators to search the outskirts of our solar system for Kuiper Belt objects (KBOs). Most noteworthy, Brown and his colleagues used images from the 48-inch taken in 2003 to discover Eris, a ball of icy rock that they later determined to be at least 25 percent more massive than Pluto, thus indirectly relegating the former ninth planet to its current status as a "dwarf planet." At the farthest point of its orbit, Eris is nearly 100 times farther from the Sun than is Earth, and it also has its own tiny moon, named Dysnomia.

With large areas of its massive steel tube covered in silver insulating foil, and an unruly tussle of wires emerging from one side, the 48-inch telescope does not have the same regal appearance it did as the backdrop for one of the more famous photos of the pipe-smoking Edwin Hubble, but it has an unassailable place in any catalog of the legendary machines of astronomy next to its giant cousin on Palomar.

The original telescope on the mountain, an 18-inch (0.45-meter) reflector, began life most famously as a tool for the ingenious and eccentric astronomer Fritz Zwicky to search for a recently discovered class of powerful exploding stars newly known as supernovae; Zwicky used the 18-inch and later other telescopes at Palomar to discover 120 supernovae, still the record for one individual. Supernovae are now considered critical distance markers in ongoing efforts to establish the scale and expansion rate of the Universe.

The 18-inch was also used to discover dozens of asteroids and comets, including the Shoemaker-Levy 9 comet, detected a year before its broken pieces slammed into Jupiter in 1994, an amazing event that spawned a worldwide observing campaign and one of the earliest media frenzies on the World Wide Web.

For the Public

Palomar Mountain (*palomar* means "pigeon haven" in La Jolla Indian) is located between San Diego and Los Angeles. To reach Palomar from these cities, exit Interstate 15 at State Highway 76 eastbound and take it for 25 miles to twisty County Road S-6, known formerly as the "Highway to the Stars" until the National Park Service tired of replacing the much-coveted road signs.

The Palomar Observatory is open daily, except December 24 and 25, from 9:00 A.M. until 4:00 P.M. The last bit of the drive runs past Palomar Mountain School, which has anywhere from one to a handful of graduates per year, mostly the children of observatory or park service resident employees.

The observatory's small visitor center was constructed in 1940. Until 2003 there was little change from the original exhibits that were installed shortly after the Hale Telescope began operations. Since then, the two dozen back-lit displays have been updated to reflect some of the historical discoveries made at Palomar and the research taking place today. The text of these displays is available in both English and Spanish. The visitor center also includes a hands-on exhibit of astronomical spectra (explaining how spectrographs break up incoming light into its component colors), several touch-screen computers with information, movies and photos from Palomar, and a large-screen display with updates from the larger world of astronomy.

Palomar's first telescope, the 18-inch (0.45-meter) Schmidt, is scheduled to be moved into an exhibit at the visitor center with information on the astronomers who have used it and their discoveries—a research history spanning more than sixty years. The move will allow it to receive the frequent attention that the first, though no longer used, instrument at this historic site deserves.

The visitor center also includes a small gift shop that is open daily during the summer months but only on weekends and holidays the remainder of the year. Unfortunately, daily visitors to Palomar are not allowed access to either the 48-inch (1.2-meter) telescope or the other medium-sized telescope on the mountain, a 60-inch (1.5-meter) commissioned in 1970 that—like the 48-inch—is now automated for remote use by observers at distant institutions.

Visitors can take the self-guided tour of the observatory that leads from the visitor center past the Palomar Testbed Interferometer (a specialized instrument, analogous to the Very Large Array for radio astronomy in New Mexico but operating at much shorter optical wavelengths, to carry out optical and near-infrared observations using ultra-high spatial resolution) and into the dome of the Hale Telescope. Five video kiosks scattered around the walking tour inside the 200-inch dome and elsewhere offer some context, images, recent scientific discoveries, and background information. A ten-minute downloadable podcast for the tour is available on the observatory's Web site.

The visitor gallery inside the 200-inch dome, reached by a long, steep stairway, provides an awesome view of the Hale Telescope, muted only slightly by the constricted feeling of being inside an enclosed glass box. Additional displays inside the large dome highlight the history of the building of the 200-inch telescope, current observing programs, and more. Handicapped access is available at the 200-inch dome at 9:30 A.M., 1:00 P.M., and 3:45 P.M.

The general public may experience a tour of the Hale Telescope without an appointment on Saturdays from April through October. Tour tickets are sold in the gift shop the day of the tour on a first-come,

The 60-inch dome with cirrus clouds.

first-served basis. No prior reservations are taken. Each tour is limited to twenty-five people each. Please note that tours may be canceled due to observatory operations or available staffing. Check the Palomar Observatory Web site for updates.

The Friends of Palomar Observatory organization supports the educational mission of the observatory by offering free access to regular tours, special behind-the-scenes tours, evening viewing sessions, and a quarterly newsletter for its members. Plans are in the works for a dedicated public outreach telescope that would allow the observatory to greatly expand its evening sessions for the Friends program and the general public.

For Teachers and Students

Behind-the-scenes tours are available by appointment year-round for educational assemblies (school groups, Scouts, astronomy clubs, and similar organizations). Local residents are generally known to have taken one tour of the observatory in middle school, prompting the frequently overheard line, "I haven't been here since sixth grade." Details are available on the Web site.

Palomar Public Affairs Coordinator Scott Kardel is available as a speaker to astronomy clubs and civic groups on the topics of light pollution, current research at Palomar Observatory, and the history of the Palomar Observatory.

A Talk with Chuck Steidel

California Institute of Technology

Chuck Steidel is the Lee A. DuBridge Professor of Astronomy at the California Institute of Technology (Caltech), an institution intertwined with Palomar Observatory more than any other over the past seven decades. Steidel earned his PhD at Caltech in 1990 and currently serves as its executive officer for astronomy. He has observed with the 200-inch Hale telescope at Palomar as much as any astronomer of his generation, and still gets a thrill when he sees it. Steidel and his collaborators use the Hale and other large telescopes, like the two 10-meter (394-inch) telescopes at Keck Observatory on Mauna Kea in Hawaii, to study how galaxies form and evolve, and he is part of a scientific advisory committee charged with outlining the highest-priority scientific uses for a 30-meter (or 98-foot!) telescope to be built during the coming decade, likely in Hawaii or Chile.

What was your first exposure to Palomar Observatory?

I started my astronomical career at Palomar as a graduate student and did my PhD thesis on the properties of gas between our galaxy and very

Chuck Steidel, California Institute of Technology.

distant quasars using the 200-inch, so the telescope is very near and dear to me. Even having used other, more modern facilities like Keck, it is still the telescope that I consider to be the pinnacle of all astronomical instruments. I'd guess that I've spent somewhere around 250 nights observing there—almost a year of my life!

I saw it for the first time as a prospective graduate student in 1984. I grew up in the Northeast and had never seen any observatory close to the size of Palomar. When I visited Caltech, I was given keys to a department car and a map to get to Palomar, about two and a half hours away. Observing on the telescope that night were two British astronomers, one of whom has become one of my closest collaborators. It was certainly that trip which made me come to Caltech, and to become an observational astronomer and not a theorist.

The 200-inch was very interesting, because it looked like a historical place and yet bolted onto the end of the telescope was the most modern possible scientific instrument. I just thought it was amazing.

When you see the thing slew around inside this gigantic dome, there's nothing like it—it looks like a battleship. It's really a home away from home for graduate students, and I still enjoy it to this day when I can get down there.

Where does the 200-inch rank today in terms of scientific power?

The very, very best nights at Palomar are about equivalent to an average night on Mauna Kea [at Keck], but from June to September, you can count on the weather being good—just when it is predictably bad in places like Arizona because of the monsoon weather pattern. There's no question that the weather is not the very best nor is the "seeing" [the image quality, determined by turbulence in the upper atmosphere] the very best, but it is very good, and it is a lot more convenient.

Palomar was "everything" when I was a graduate student—a big telescope with a powerful spectrograph. We are still using Palomar as a wide-field imager to feed spectroscopic observations at Keck, much the same as the 48-inch on Palomar once fed targets to the 200-inch back in the day. It is a near-perfect complement. The 200-inch also has a very good adaptive-optics system [where a laser beam provides a guide-star target to correct for much of the distortion caused by Earth's atmosphere]. Palomar's got quite a few state-of-the-art instruments, with a few more new ones to arrive in the next year or two.

Caltech has a smaller share of observing time now on the 200-inch than we do on Keck, so it tends to be just as oversubscribed [by astronomers proposing to use it] as Keck. It's a great place for experimental instruments and for students to learn how to observe. We know we're not going to use it forever—sometimes you have to give up things to get something new—but it's still quite important and there's still a large demand for it.

Palomar is considered the original buttoned-down bastion of astronomy, a comparatively regal place during its first several decades. What is the general environment like today?

It's still operated basically the same way it's always been operated. One thing that's very special is that some of the people working there are the same people who have been there since the first time I went there, so it's almost like seeing old family members. I know that's quite unusual. One man was a night assistant on the 200-inch for about thirty-five years—he was a total "fixture" who worked with all the greats. I've been up there when there were large snowstorms, and the major forest fires in 2007 came within about a mile of the summit.

Another unique thing is Mother's Kitchen, which is a vegetarian restaurant on the way up that is often frequented by big guys on motorcycles who like to ride through this beautiful forest. Unfortunately, the trees have been really devastated by the bark beetle in the last few years. There's been a huge effort in the national forest to clear it out; it will take some time for the trees to come back. There are now housing developments for about two-thirds of the drive down there, and huge casinos where there used to be small Native American villages. Once you get past that, it looks pretty much the way it always has, with beautiful orange groves. I really enjoy the final 20 or 25 miles of the drive.

The place where astronomers stay is referred to as the Monastery. It's basically a house with bedrooms, but the food is still cooked and served family style, by people who really care, and it's still very good. It's like having a mom down there. It doesn't have all the modern conveniences, but it's a really nice place to stay.

What connections do you see between the 200-inch and the next generation of large observatories like the Thirty Meter Telescope (TMT) project?

One big reason that Caltech is involved in the TMT is that we have a long history of large telescope projects. Hale was one of the founders of Caltech itself, and the beginnings of Caltech more or less correspond to the start of the 200-inch, when Caltech was the lucky recipient of a Rockefeller Foundation grant. We later partnered with the University of California to do the Keck telescopes. It was actually the Caltech administration, not the astronomers, who asked us about ten years ago to start thinking about "what next?" It's pretty unusual for an institution to be so committed to astrophysics, and the TMT will be a big part of our future.

The reason the 200-inch lasted so long is that all kinds of huge gains could be made with improved instruments and technology, like the switch from photographic plates to digital detectors. Now we're at the point where the only way to get really new capabilities is to think about larger apertures. That's why there's been a relatively short period between when Keck was completed and when we started thinking about TMT.

No matter where the TMT ends up being sited, it will be in a much more remote and hostile environment than Palomar, where it will be unlikely that many people will travel to use it after its initial years, particularly if it ends up in the Southern Hemisphere. There will be much less of a direct relationship with the "place." It will be a newfangled device, but it certainly won't have the charm and tangible history of Palomar.

How will history regard the 200-inch telescope? What should people take away from a visit?

The thing about the 200-inch that I find really amazing is that, since they did not have computers back then, the telescope (and its pointing capabilities) had to be all mechanical, so it was sort of overdesigned. This makes it much less likely to break than more modern telescopes, and it has a lot fewer problems with interactions between the hardware and software. I've lost almost no observing time at Palomar due to technical problems, which I can't say about too many places.

The people who work at the place really, really care about it—perhaps even more than is healthy! The dome floor is always immaculate. Most of the staff are there because they love the place, and there really is no substitute for that.

The 200-inch is an example of people thinking really big and doing something that they really did not know how to do when they started. It was a tough time to do anything like it during the Great Depression. It provided jobs for a lot of skilled people who happened to be out of work, which is one reason it turned out so well. People were incredibly excited about the mirror when it traveled across the country. It's interesting to recall what a huge cultural icon it was when it was completed, and the telescope is still doing good science today, sixty years later. It's an amazing place to visit.

Science Highlight

The Cores of Active Galaxies

In the early 1960s, the Palomar 200-inch telescope was used to carry out a study of the starlike optical counterparts of mysterious sources discovered in the course of a deep survey of the sky with radio telescope dishes carried out in Cambridge, England. In 1963 and 1964, astronomers Maarten Schmidt and Jesse Greenstein used the enormous light-gathering power of the 200-inch telescope to record photographic spectra (the component colors) of the optically brightest counterparts to these radio-wave sources.

What they found after examining their eight-hour-long exposures was truly remarkable: Objects that exhibited emission lines of oxygen and nitrogen (among other elements) shifted to far longer wavelengths of light (redward) as compared to the "rest" wavelengths of the elements as measured in the calm conditions of terrestrial laboratories. A "redshift" of this observed magnitude implied that these objects were receding from Earth at velocities in excess of 100,000 kilometers per second. If these apparently starlike objects were galaxies, they would lie at distances exceeding 3,000 million light-years. When placed at that vast distance, because absolute brightness is proportional to apparent brightness times distance squared, their luminosity must exceed the brightness of the Milky Way by a factor of nearly 100 times.

Were these objects galaxies? Were the observed redshifts truly indicative of their distance or some trick played on light caused by local forces? If they were such distant galaxies, why did they appear starlike? What might account for their enormous brightness?

Astronomers at Palomar and throughout the world set out to uncover the nature of these mysterious objects. A steady trickle of new data began to tell the story of these so-called quasi-stellar objects (QSOs). One key element of the story was the discovery—both at Palomar and at Kitt Peak National Observatory—of a suite of absorption lines arising from both hydrogen and heavier elements (carbon, oxygen, silicon) superimposed on the bright, continuous spectrum of the QSO. These absorption lines also showed redshifts, but had values smaller than the redshifts of the emission lines arising from the QSO.

These newly observed absorption lines seemed best explained by assuming that they arise in intergalactic gas lying at a variety of distances between Earth and the QSO (also known as a "quasar") under study. Second, astronomers at Kitt Peak and Mount Hopkins discovered a few cases of "double QSOs" located in the vicinity of a bright, relatively nearby galaxy. Spectra of each member of the pair revealed that they had identical redshifts—a result that seemed best explained by postulating that the two objects were actually images of a single QSO located behind the nearby galaxy, but "lensed" into two images by the powerful gravitational field of the intervening galaxy, an effect predicted by Einstein's general theory of relativity. The presence of intervening intergalactic absorption lines as well as gravitational lensing by an intervening galaxy provided strong evidence that QSOs were indeed located at the distances implied by the redshifts of their emission lines.

The fact that QSOs are galaxies came via the discovery of faint halos of light surrounding their bright stellar-like "cores." Spectroscopy of their halo-like "fuzz" revealed absorption lines sharing the same redshift as the starlike core and with a pattern of line strengths characteristic of those found from spectroscopic studies of nearby galaxies. Such a pattern was consistent with what could be expected from the combined light of billions of stars.

During the 1960s and 1970s, astronomers at Palomar and elsewhere, including Maarten Schmidt and Wallace Sargent, were able to show that the patterns of emission-line strengths exhibited by QSOs were similar in many respects to those shown by the bright nuclear regions of galaxies located far closer to Earth than the distant QSOs. Many of these closer systems also were active emitters of radio waves, strengthening the possibility that these galaxies with "active galactic nuclei" (AGN) might be cosmic cousins of QSOs. Together, this evidence suggested that QSOs might be an unusual and rare class of galaxies, characterized by strong radio emission, extraordinarily luminous starlike cores, and "halos" comprised of stars, all within a volume comparable to that of the Milky Way.

The proximity of nearby AGN-type galaxies allowed detailed study of the physical conditions that characterize their bright nuclei. Astronomers have combined optical and radio observations of AGN and QSOs to develop models of increasing complexity and fidelity that seem to explain the optical and radio properties of the nuclear regions of AGN galaxies and QSOs. These models suggest that both the optical and radio emissions ultimately arise from processes that occur when gaseous material surrounding the galactic nucleus accretes onto a massive black hole located at the galaxy's center.

Accretion onto the black hole appears to be fed by a swirling gaseous disk, threaded by magnetic fields. Most of the gaseous material spirals inward through the accretion disk and ultimately is gobbled up by the black hole. Along the way, the gas is heated and releases energy in proportion to the accretion rate and the mass of the black hole. This emission emerges as highly concentrated, powerful optical and x-ray emission, which has been detected by orbiting x-ray observatories such as NASA's Chandra X-Ray Observatory. (This is why most astronomers who study AGN use data both from the ground and from space.)

A small fraction (perhaps 10 percent) of the accreting material from the disk surrounding the black hole appears to be ejected in spectacular, highly collimated "jets" of ionized plasma. As the charged particles in the plasma interact with the magnetic field of the jet, they produce strong radio output of a particular type known as "synchrotron emission." QSOs are characterized by highly luminous nuclear emissions produced as black holes with very large masses (millions of times the mass of the Sun) are fed by surrounding disks of gaseous material that accretes onto the black hole at high rates. Typical nearby galaxies hosting AGN have lower, though still powerful, emission from their nuclei—a result of lower black hole masses and/or lower gas-accretion rates.

Following the discovery of the first QSOs, astronomers using Palomar telescopes carried out surveys aimed at detecting large samples of objects so that they could begin to use statistics to uncover the range of properties of these objects, as well as to chart their evolution over the lifetime of the observable Universe. Key to these studies was the observation that many QSOs showed unusually blue colors—in analogy with the observed colors of nearby AGN (a result of emission arising from heated gas in the accretion disk surrounding the black hole).

Maarten Schmidt and Richard Green used the wide-field 48-inch Oschin Telescope at Palomar to search the sky for unusually blue objects with bright nuclei by taking photographs through a set of colored filters designed to enable separation of QSO and AGN candidates from the billions of ordinary stars that show up on such photographs. This survey, combined with follow-up observations carried out at observatories throughout the world, led to the discovery that the fraction of QSOs as compared to normal galaxies seemed to be higher when the Universe was young—at an age of 1 billion to 2 billion years or around 10 percent of the current estimated age of the Universe, about 14 billion years.

Astronomers are still trying to fully understand what combination of properties (for example, black hole mass, gas-accretion rate, location of galaxies in different environments) might account for this evolutionary pattern. Current thinking suggests that all relatively massive galaxies—galaxies similar to and more massive than the Milky Way—develop or are born with a central black hole and may go through several "active" phases when material accretes onto the black hole. Perhaps the greater abundance of QSOs at early epochs reflects the period when black hole masses are building and the supply of surrounding gas available for accretion onto the black hole is still high. At later epochs, though the black hole masses have reached their limits, perhaps the gaseous "fuel" for feeding accretion disks is depleted. Current forefront research attempts to understand how the black holes at the center of galaxies are born and what role these black holes, their accreting gas, and their mass outflows play in affecting the evolution of the surrounding galaxy.

The presence of extraordinarily bright QSO "beacons" that can be observed to enormous distances provides astronomers with a tool to chart the spatial distribution of intergalactic gas and to map the rate at which elements heavier than hydrogen are produced and injected into the intergalactic medium. Wallace Sargent and later Chuck Steidel have made use of QSOs as background sources against which they have observed the redshifts and strengths of absorption lines produced as light from quasars passes through colder intergalactic gas located between Earth and the quasars. By observing the spectra of a large numbers of QSOs, and isolating absorption lines arising at the same redshift, Sargent, Steidel, and others have begun to construct a three-dimensional picture of the distribution of intergalactic gas.

This picture reveals a weblike distribution similar to that discovered in redshift surveys of galaxies at Mount Hopkins (see pages 90–91) and elsewhere. Eventually, when a full three-dimensional map is extended to larger redshifts and sufficiently large volumes, it should be possible to understand the relationship between this weblike structure and the density fluctuations imprinted at the time of the Big Bang (the cosmic explosion that gave birth to the Universe more than 13 billion years ago). Astronomers have compared the strength of absorption lines arising from elements heavier than hydrogen in intergalactic gas located at different distances from Earth; the more distant gas provides a glimpse into the conditions that obtained when the Universe was younger. By observing absorption line strengths in more distant (younger) gas and in the nearby intergalactic medium, astronomers have been able to track how the abundance of different elements increase as the Universe ages. This increase matches the concept that supernovae explosions and other evolving massive stars produce these heavy elements and subsequently eject them into the medium between galaxies. (For more on this, see page 147.)

Indeed, more than 90 percent of the mass of elements in our human bodies originated as such "star stuff," giving us all a stake in a better understanding of this amazing process.

The exterior of the eighteen-story-tall Mayall Telescope.

Kitt Peak National Observatory

Land of the Tohono O'odham Nation, near Tucson, Arizona

The World's Largest Collection of Research Telescopes

The largest collection of research telescopes on any mountain in the world can be found in southern Arizona near Tucson, perched on Native American land atop the highest peak in the Quinlan Mountains.

Kitt Peak National Observatory has been a special place for astronomers and visitors from all over the world since it was founded a half-century ago. Twenty-three optical telescopes and two radio telescopes are sprinkled across the ridges of the mountain, amid a landscape of piñon pines and silverleaf oak trees. Located at a relatively modest altitude of 6,875 feet, dry desert vistas run off in every direction from Kitt Peak. This tranquil landscape also continues to be the home of a wide variety of desert plants, including cacti, and wildlife, such as coatimundi, lizards, swarms of ladybugs at certain times of year, and omnipresent turkey vultures.

Web site
www.noao.edu/outreach/kpvc

Phone
(520) 318-8726

E-mail
outreach@noao.edu

Address
55 miles southwest of Tucson, Arizona, on AZ Route 86/386

The largest telescopes on Kitt Peak National Observatory are funded primarily by the National Science Foundation (NSF). These telescopes—the Mayall 4-meter (158-inch), the 2.1-meter (84-inch), and part of the 3.5-meter (138-inch) WIYN Telescope—comprise a larger organization called the National Optical Astronomy Observatory (NOAO). The WIYN Telescope is so named because it is operated jointly by the University of Wisconsin, Indiana University, Yale University, and NOAO. In addition to Kitt Peak, NOAO operates a similar observatory in South America, called the Cerro Tololo Inter-American Observatory (CTIO), located in northern Chile at the jagged edge of the Andes, thus giving U.S. astronomers access to the wonders of the southern sky, including the center of our galaxy.

The three large telescopes on Kitt Peak that continue to be operated by NOAO are open to competitive access, meaning that, every six months, committees of astronomers review proposals from their peers throughout the United States. The proposals are matched against the available telescopes, the instruments on them, and the number of nights requested, and the proposals judged the very best are awarded valuable telescope time. This provides nationwide access to major observing facilities (hence the name "national observatory") and helps ensure that the very latest and most creative ideas in astronomy can be explored and developed—weather permitting!

Most of the other telescopes on Kitt Peak are owned and operated by university groups or similar consortia, as well as the National Radio Astronomy Observatory (NRAO), the national counterpart to NOAO for wavelengths of light much longer than optical or infrared waves (see related material on New Mexico's Very Large Array, pages 93–98).

The roots of Kitt Peak as a center of astronomical research date back to August 1953, when thirty-five astronomers gathered at Lowell Observatory in Flagstaff, Arizona, for a meeting sponsored by the fledging NSF. The meeting's agenda focused on a proposal by three universities—the University of Arizona, Ohio State University, and the University of Indiana—to jointly build and operate a new telescope. At that time, research-quality telescopes belonged mainly to large private institutions, and astronomers who worked for smaller institutions did not have equal access to state-of-the-art equipment or, in some cases, any telescope at all.

Among the attendees at that historic 1953 meeting was Leo Goldberg of the University of Michigan. Goldberg's active interest in the future of U.S. astronomy led him to propose the founding of a national observatory that would offer the finest research instruments to the country's astronomers on a competitive basis. The resolution won NSF endorsement and a panel was convened, chaired by Robert McMath. McMath was an industrialist and engineer who took up astronomy as a hobby and years later joined the ranks of academia at the University of Michigan, eventually becoming the first president of the Association of Universities for Research in Astronomy (AURA), the nonprofit consortium of thirty-one universities that continues to manage Kitt Peak National Observatory and NOAO.

The panel recommended that the NSF support the construction of a national observatory equipped initially with 36-inch (0.9-meter) and 84-inch (2.1-meter) telescopes to observe objects ranging from denizens of our solar system to the most distant galaxies, as well as the world's largest telescope devoted to detailed study of the Sun.

But before an observatory could be built, a suitable location had to be found. Observatories have unique requirements, including clear and dry atmospheric conditions around and above the site and minimal light pollution nearby. After a search covering thousands of square miles, the search was narrowed to 150 sites and, ultimately, five prime candidates, four in Arizona and one in California. Further stringent comparisons narrowed the choice to Hualapai Mountain near Kingman, Arizona, or Kitt Peak.

Intense site testing was initiated to see which of the two would best match the exacting criteria demanded by the astronomical community. Concerns such as the amount of developable area and the amount of local foliage, proximity to a major university and support facilities, airflow, and transparency of the sky were decided relatively quickly. Other features called for specialized testing. A 16-inch (0.4-meter) telescope was installed on each site to test both for the transparency of the atmosphere and for its expected "seeing" conditions, a term used by astronomers to describe the stability of the atmosphere and the resulting clarity and stability of astronomical images.

After three years of rigorous testing, Kitt Peak emerged as the choice.

Yet there was one more major step to take in the process of securing Kitt Peak as the site for the national observatory.

Kitt Peak is located on the land of the Tohono O'odham Nation, the second-largest Native American reservation by area in the United States. Moreover, the mountain, called "Ioligam" by the O'odham (which translates to "mountain of manzanita wood"), is a sacred place in tribal lore. The mountain and the surrounding land are considered to be the garden of I'itoi, the big brother spirit that taught the indigenous people to become self-sufficient in the desert (the name *Tohono O'odham* means "People of the Desert"). Some of the Tohono O'odham people, who were then known in the Anglo community by the colloquial name Papago, feared that the proposed observatory would disturb the tranquility of Ioligam and ruin its natural beauty. It was not until Edwin Carpenter of the University of Arizona's Steward Observatory invited the tribal council to the campus to view the night sky as astronomers view it—through a telescope—that most council members became convinced that the observatory could exist in harmony with their sacred mountain.

Permission was granted, and a lease agreement was drafted and formalized by Congress. Signed in October 1958, the lease granted 200 acres of the mountaintop for scientific use and an additional 2,200 acres as a buffer zone, in perpetuity as long as the lease site is used for science. In return, AURA provides an annual rental fee to the Nation, unfettered access to the site for tribal visits and rituals, and free access to its public outreach programs.

Four 16-inch telescopes would be in place for research and testing on the mountain by 1964. But the first of the large telescopes installed at Kitt Peak was the "Number-One" 36-inch, completed in March 1960. That telescope was located on the spot occupied today by the WIYN 3.5-meter telescope. The "Number One" 36-inch telescope can now be found at a site up the hill west of the Kitt Peak Visitor Center and is operated by the Southeastern Association for Research in Astronomy (SARA), a consortium of universities and colleges located in the U.S. Southeast. The "Number-2" 36-inch at Kitt Peak followed in 1966. Today, this telescope is operated by the WIYN consortium.

The 2.1-meter telescope, the second major telescope installed at Kitt Peak, employed major advances in mirror-making techniques.

Manufactured by Corning Glass Works, its 3,000-pound mirror was fabricated by a process called sagging, in which large chunks of Pyrex glass were placed in a mold and then heated to the melting point. As the glass melted, it filled the mold. Inside the mold were cores around which the glass flowed to create a honeycomb effect in the back of the mirror. Prior to this, molten glass was ladled into molds, a process that resulted in greater imperfections within the mirror and higher production costs. The new process, though not flawless, was successful, and the 2.1-meter telescope opened in September 1964.

Early in its research life, it was used to discover the Lyman-alpha forest, a series of spectral absorption lines that appear in the shorter wavelengths of the spectra of quasars—produced by elements populating the gaseous medium between Earth and these distant galaxies. Measurements with the 2.1-meter telescope verified the phenomenon of gravitational lensing—predicted by Einstein's general theory of relativity—in which galaxies act as lenses to bend and redirect the light from more distant objects that lie behind them in the direct line of sight.

Despite its age, the 2.1-meter telescope still attracts large numbers of highly competitive proposals for observing time and continues to be the source of significant astronomical discoveries. It also serves as the proving ground for new generations of telescope cameras and spectrographs that observe at infrared wavelengths, which are longer than the human eye can see.

Just a half-year after the 2.1-meter mirror blank left Corning Glass Works for its trip to Tucson, ground was broken for what would become the world's largest solar telescope, a title it retains to this day. Awesome in both design and dimension, the 1.6-meter (63-inch) McMath Solar Telescope required workers to build not just on the mountain but *into* it as well. The vertical tower that supports the three sun-reflecting mirrors, or heliostats, rises 100 feet above ground level. The diagonal light path, aligned with Earth's polar axis, spans 200 feet from the top of the heliostat tower to the ground and then continues another 300 feet belowground through the mountain's granite interior. The McMath Solar Telescope Facility was dedicated in November 1962, at a development and construction cost of nearly $4 million. On its thirtieth anniversary in 1992, the telescope was rededicated as the McMath-Pierce Solar

Panorama of Kitt Peak National Observatory with the McMath-Pierce Solar Telescope in the foreground.

Telescope, adding the name of astronomer Keith Pierce, who was integral to the telescope's design and implementation under original patron Robert McMath. Pierce used the facility to carry out a range of important scientific investigations over a period of nearly five decades.

The McMath-Pierce Solar Telescope is unique for its liquid-cooled exterior. The structure's outer skin consists of 140 copper panels that contain 18,000 gallons of antifreeze solution, which can be circulated to help maintain a desired temperature. In addition, the tunnel portion uses a refrigerated liner requiring about 25,000 feet of one-inch pipe to maintain a stable column of air in the optical path during the winter months, when the air inside the telescope, absent cooling, would be warmer than the air outside. Rising warm air would distort the image of the Sun. Cooling is thus critical to producing ultra-sharp images of the Sun's surface.

Among solar telescopes today, the McMath-Pierce remains special for its ability to observe the Sun in infrared wavelengths. This is accomplished through the use of an open-air reflecting optical system that never passes the Sun's light through anything that might be opaque to infrared light. This adaptability has kept the McMath-Pierce in step with changes in the research challenges of solar physics and at the forefront of solar observing. Its scientific legacy includes the discovery of water molecules forming over the coolest areas of sunspot umbrae and mapping of the sodium emission from Mercury.

In March 1968, construction began on what many consider the crown jewel of Kitt Peak—the 4-meter Mayall Telescope. The first step in the actual construction of this imposing eighteen-story facility was pouring its tall central pier. Because of ground turbulence created by wind flow over the summit, the telescope needed to be raised considerably above ground level so that this turbulent air would not distort the incoming starlight.

A hollow pier 92 feet tall and 37 feet in diameter was designed and then constructed by a process involving a slip form. A circular form surrounded by scaffolding and cranes was placed at the bedrock and filled with concrete. As the concrete at the bottom of the form solidified, the form was raised and more concrete was poured into it, and so the process continued under blazing artificial light for three consecutive days and nights.

When the monolithic pier stood firmly in place, its 18-inch-thick walls reinforced with steel, workers began constructing the observatory dome and exterior infrastructure around it. To avoid the possibility of any vibrations being transmitted to the telescope, the building and the pier are mechanically isolated from each other. The next phase of construction involved placing the hexahedron-shaped supports around the pier. The giant 96-foot-tall structural supports were fabricated in Phoenix and trucked to Kitt Peak fully assembled, then installed in place at the rate of one per day. The building required ten of these 35-ton units.

A 500-ton dome was placed atop the building, designed to withstand winds of 120 miles per hour. The dome, which can rotate completely in less than six minutes, features a 27-foot-wide shutter weighing 30,000 pounds that can be opened or closed in two and a half minutes.

With a building this large, it was possible for designers to incorporate features not usually found in smaller observatories. Inside the structure on floors below the telescope are living quarters, complete with kitchen and showers, a lounge and recreation room, and, typical for telescopes of the era, darkrooms for developing the glass plates used for photographing astronomical objects. Unfortunately, even the tiny amount of heat given off by visiting astronomers (which might migrate into the dome in some small way and, again, cause air currents that would blur telescope images) was worry enough to move visitors to a central dorm on the mountain.

If the observatory for the 4-meter telescope seems imposing, the telescope is well matched to its surroundings. The mirror alone, fabricated with quartz glass because of its lower susceptibility to thermal changes than Pyrex (the glass used for the 2.1-meter Kitt Peak telescope and the 200-inch Palomar telescope), weighs a staggering 15 tons. The contract for the mirror was awarded to General Electric in 1964. It was 1967 before the finished blank, costing $1,150,000, arrived in Tucson for grinding and polishing. That process would take an additional three years and require the designing of a special polishing machine to accommodate the large mirror. When the polishing was completed, the glass disk was covered with a coating of aluminum just one-thousandth the thickness of a human hair. It is this thin coating of aluminum that reflects the light from distant astronomical objects to other optical elements and ultimately to instruments that record and analyze the light from cosmic sources.

The 4-meter telescope, like the 2.1-meter and many other smaller telescopes on Kitt Peak and elsewhere, uses an equatorial mount, so named because this type of mount is tipped at an angle to make it aligned with Earth's rotational axis. That alignment allows the telescope to move in an arc, paralleling the way the stars move across the sky, and prevents objects from changing their orientation in the field of view; this allows telescope operators to track stellar objects by rotating the telescope around a single axis, rather than needing to perform a complex series of tiny up-and-down pointing movements in two directions.

But to support a telescope mirror as massive as the 4-meter's, an equatorial mount must be extraordinarily large and robust—the moving parts of the 4-meter telescope weigh about 300 tons. The horseshoe that supports the telescope, similar to the one used on the Palomar 200-inch telescope in California (see pages 7–8), turns on a thin film of oil about four-thousandths of an inch thick. Despite their bulk, the mounting and telescope are so perfectly balanced that they are moved by a mere half-horsepower electric motor.

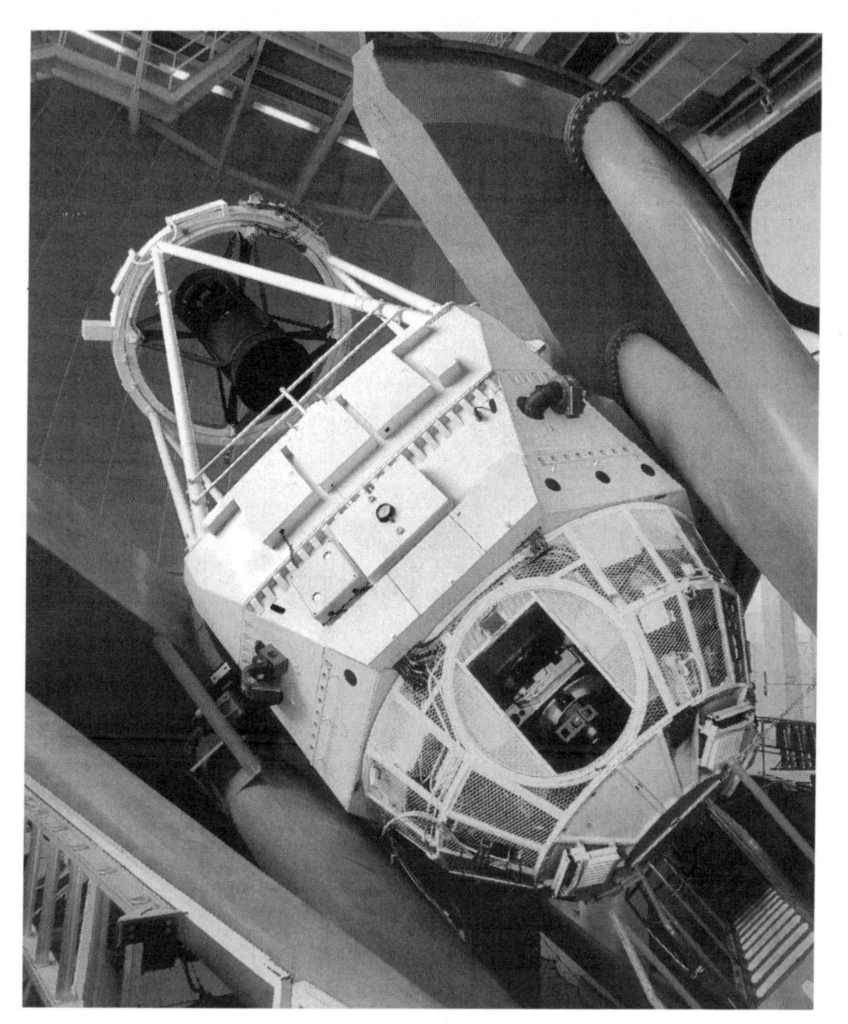

The 4-meter Mayall Telescope, dedicated in 1973.

Outfitted with other recent improvements such as exterior dome vents to maintain steady temperatures and airflow, the Mayall Telescope continues to do its job well today, for both individual observers and large teams of astronomers. New and more powerful instruments are now being built in partnerships with several universities.

If the 4-meter represents the best of an earlier generation of technology, the WIYN Telescope represents the modern concept of telescope design. With a primary mirror diameter of 3.5 meters, the WIYN Telescope is only marginally smaller in aperture than the 4-meter telescope. Yet, the differences in size and function are remarkable.

The WIYN employs a relatively thin mirror fabricated by a method called spin casting. In this process, borosilicate material is melted into a mold that spins at seven revolutions per minute. The spinning causes the molten glass to flow up the sides of the mold and give the mirror its basic configuration. This process results in a thinner and therefore lighter mirror. The 3.5-meter mirror weighs 4,000 pounds, about one-seventh the weight of the 4-meter mirror. It is very similar to the 3.5-meter primary mirror at Apache Point on Sacramento Peak in New Mexico (see pages 118–119).

The equatorial mounting used on the 4-meter and 2.1-meter telescopes gave way to a much more compact and simpler type of mounting called an alt-azimuth mount. In this design, the telescope sits on a circular base with two upright supports rising on either side, which each hold large bearings that cradle the telescope. Thanks to computer technology, the telescope can be tilted up and down (altitude) on these bearings and pivot on the circular base horizontally (azimuth) at the same time, in such a controlled way as to precisely track the movement of objects across the sky. The final result is that the moving parts of WIYN weigh only 40 tons.

Kitt Peak offers a self-contained community in support of its visiting scientists. The cafeteria is accessible twenty-four hours a day. Four dormitories are home to visiting astronomers and mountain staff where they can sleep during the day after completing their night's work at the telescopes.

Today, Kitt Peak is operating new instruments for wide-field digital imaging at the optical and infrared wavelengths, thus contributing a key element to the national, multi-wavelength "system" of astronomical facilities designed to offer the maximum benefit to the entire community of astronomers in the country. Its dark and clear skies remain among the best in the continental United States, and there is a strong consensus among astronomers to keep Kitt Peak open and on the cutting edge of science for decades to come.

The 3.5-meter WIYN Telescope, showing a spectrograph (black disk) mounted to its dedicated Nasmyth instrument port, which is fed light from stars and galaxies via optical fibers inside the silver tubing.

For the Public

Astronomers are not the only ones who are invited to Kitt Peak. Funding from the National Science Foundation carries a mandate to promote public understanding of science and support for the value of basic research.

The Kitt Peak National Observatory Visitor Center is open to the public daily, with the exception of Thanksgiving, Christmas, and New Year's Day. The observatory is located 55 miles southwest of Tucson on the Tohono O'odham reservation, about a 75-minute drive from downtown, traveling from Interstate 10 to Interstate 19 South and then west out the scenic Ajo Highway (Route 86) to Route 386. A 12-mile drive up one of the more gentle roads in the world of astronomy will take twenty minutes after the turn. Count on spending about four to five hours round-trip for the drive to the mountain and a docent-guided tour of one of the major telescopes. Be aware that Kitt Peak is at an altitude of nearly 7,000 feet and that some of the hills leading up to the facilities demand a fairly steep walk.

Kitt Peak offers a variety of public programs through its visitor center, as well as a gift shop featuring books, clothing, observing aids, and a widely known collection of Tohono O'odham baskets. The visitor center is free, but there is a small fee for each of the three docent-led daily tours, each of which covers one of the major telescopes: the Mayall 4-meter, the 2.1-meter, and the McMath-Pierce.

The NOAO public outreach department also offers occasional special classes in CCD (charge-coupled device) imaging, meteor shower observing, and how to get the most out of your new telescope, as well as occasional cultural activities like "Stars & Music" events that blend stargazing with the performing arts.

For people who would like to experience the thrill of observing, the visitor center offers a widely known Nightly Observing Program. Designed to introduce the public to the wonders of the night sky, this paid four-hour program features instruction on the use of star charts and binoculars, an opportunity to use the charts to identify constellations

Kitt Peak Visitor Center with 20-inch public telescope dome.

and bright stars and to observe bright deep-sky objects with the binoculars, and guided observing with one of the three public telescopes, each operated by knowledgeable and friendly staff.

For those who seek an observing experience closer to that of the professional astronomers, the visitor center offers the Advanced Observing Program, in which participants are treated like visiting astronomers and given a dorm room, meals in the dining hall, and a public telescope for the night to do visual observing, astrophotography, or CCD imaging. The participant plans the run and executes it with the assistance of a telescope operator, just as a professional astronomer would do. The 20-inch telescope in the visitor center observatory used for this program is also available for remote programs over the Web, allowing anyone to observe at world-famous Kitt Peak using their own computer.

Access to all of these programs is free to the people of the Tohono O'odham Nation, with advance reservation recommended.

For Teachers and Students

NOAO offers programs for all ages, from hands-on programs for elementary school and middle school using materials such as Project ASTRO, to more advanced programs for high school and college students conducted by its unique group of astronomer-educators. NOAO also is the national headquarters for a self-contained set of hands-on kits for optics education full of mirrors, lasers, lenses, and kaleidoscopes ("Hands On Optics"), and it participates in an Astronomical Society of the Pacific program to train smaller science and nature centers in easy-to-use public programs on the Moon and stars ("Astronomy from the Ground Up"). Since 2006, NOAO has coordinated an annual worldwide program every March on dark-skies awareness known as GLOBE at Night, which encourages the public in both hemispheres to get outside and observe the constellation Orion and compare the number of stars that can be seen with templates of a truly dark, natural sky.

A Talk with Vera C. Rubin

Department of Terrestrial Magnetism, Carnegie Institution of Washington

Vera Rubin is an observational astronomer who studies the motions of gas and stars within galaxies and the motions of galaxies in the Universe. She was influential in discovering that the Universe contains invisible "dark matter," whose presence is inferred from its gravitational pull on motions of stars and galaxies. Since 1965, she has been a staff member at the Department of Terrestrial Magnetism at the Carnegie Institution in Washington, D.C. She is a member of the U.S. National Academy of Sciences, and in 1993 President Clinton awarded her the National Medal of Science.

When did you first starting using telescopes at Kitt Peak National Observatory?

In early 1963, I attended a meeting of the American Astronomical Society in Tucson and went to Kitt Peak with a group tour. At the 36-inch telescope, the largest stellar telescope operating there at the time, there were labels printed on each major piece: "Pull Here to View Object." "Look Here to See Guide Star." "Push Here to Close Exposure." I had only used student telescopes at that point, but I looked at all the labels and thought, "I can do that." So I submitted a proposal to observe stars in the opposite direction from the center of our galaxy, in order to try and map the velocity of rotation in the outer parts of the galaxy.

I started observing at Kitt Peak on the 36-inch telescope in 1963, and graduated to the 84-inch (now commonly called the 2.1-meter) in 1965. Some of my most significant work was done there. The telescopes were wonderful, the telescope operators were very helpful, and the views of the surrounding desert flowers from the location of the 36-inch on the south ridge of Kitt Peak were breathtaking. I would get to the telescope before sunset, get all the equipment prepared for the night's work, and then walk out to the west of the telescope to watch sunset.

I remember once I mentioned to the director how beautiful the site was, and he replied, "That's what you can do if you pour one million dollars into a mountaintop." But of course, the real beauty was the telescopes and the people who helped us carry out our programs.

What was involved in your work on key evidence for the existence of dark matter?

Kent Ford and I worked together using an innovative new instrument that he designed and built called an image-tube spectrograph. This device could take detailed measurements of the spectrum of light from distant objects and do it ten times faster than unaided photographic plates. Our goal was to observe faint regions in the spiral galaxy M31 (Andromeda), and measure the velocity of the rotation of these regions around the center of that galaxy to see how their speeds changed beyond the visible parts of the galaxy disk.

These faint regions of ionized hydrogen gas, which scientists call "H-alpha" regions, were mapped by Walter Baade in the 1940s using the 100-inch telescope at Mount Wilson near Los Angeles. These regions

Vera Rubin at the Kitt Peak 2.1-meter telescope.

were too faint to be visible in the 84-inch telescope at Kitt Peak or the 72-inch at Lowell Observatory, which we also used, so we had to hand-guide the telescope in the dark, using "blind offsets" measured from Baade's maps. We worked in total darkness so as not to cause stray light in the instrument. It was a hard task in a cold, dark dome with freezing fingers inside heavy gloves.

Using Newton's laws, we could use these velocity measurements to infer the mass of the galaxy. Newton's laws predict that the rotation velocities of these H-alpha regions around the center of a spiral galaxy should get slower quite rapidly as they get farther out from the center, if the distribution of mass simply follows the obvious decrease in light that we can see. So we mapped our data and found that these H-alpha regions continued to have high velocities over the entire visible part of the galaxy. This "flat rotation curve" offered strong evidence for the existence of some sort of significant gravitational pull that could not be seen. This is what we call dark matter. We still don't know what it is today.

Our work was an important chapter in the dark matter story, although the evidence was not overwhelming until work at the then-new Kitt Peak 4-meter telescope in the 1970s and early 1980s. We observed more distant galaxies, so a single exposure could cover the long axis of the galaxy. We also observed galaxies at the Cerro Tololo Blanco 4-meter telescope in Chile.

We published three major papers starting in 1978: one of rotation curves for Sc-type spiral galaxies, one for Sb types, and finally one in 1985 for Sa-type galaxies that included a comparison of the three types. Most of my studies are still concerned with motions in galaxies. As the technology continues to develop, we ask and answer even more complex questions.

What kinds of questions does dark matter raise?

It forces us to ask, "What is keeping the stars and gas spinning around so fast, beyond the part of the galaxy that we can see? What's keeping the stars that are very far from the center from flying out into space?" The current answer is, "Gravity from a type of matter that does not emit any light."

Like many advances in science, dark matter was not discovered by one experiment or one observation by one person or a team, at a single instant. It took hard work, over many years of observing by us and by others, to be confident that we understood at least some of what nature was showing us—that I was part of this chain gives me great joy.

But the story is not over. There are still too many details that we do not understand, and that is why science is truly the "endless frontier." I think that many more surprises are in store. Our children and grandchildren will build on the foundations that we have erected, and they will also correct our misconceptions and errors. Maybe they will be the first generation to answer the big question, "What is gravity?"

Do you recall any extra-special or funny moments at Kitt Peak?

When Kent and I used the Kitt Peak 84-inch telescope in the late 1960s, we trucked our heavy spectrograph and all its auxiliary equipment back and forth between Kitt Peak and Flagstaff. At the loading dock of the 84-inch, we were met by four or five Native American workers who would help us unload. I still remember one of them saying, "There must be an easier way to make a living." I never knew whether they were talking about themselves or us.

We were taking measurements of very faint regions in M31. While exposing the photographic plates at the telescope, we needed total darkness in the dome. We took very long exposures—some as long as six hours—and we wanted no stray light to leak onto the plates. The telescope operators were required to spend the night in the dome, and they would cringe when we showed up. They knew that they would have nothing to do all night, once they set the galaxy in the telescope field of view. In fact, they even covered the small clock, so the dim red light would not show. I'm not certain that they would call this "fun," but they were always good sports about it.

You have visited and used many observatories around the world. What would you say is the most distinctive thing about Kitt Peak?

From the start, Kitt Peak was very special because it would accept applications from any astronomer. Telescope time is awarded on the basis of the proposal: What observations does the astronomer want to make, and why are these observations important? Before this, telescopes were largely the property of universities or research institutions, and only their faculty or research staff had access to them. Kitt Peak and Tucson were exciting places to visit, because they became the center of the world for U.S. astronomical observers.

It was a special joy to meet astronomers I had known only by name from their scientific papers. Breakfast time, at mid-afternoon, was always a time for interesting discussions that ranged from last night's weather to tonight's weather, astronomy, our families, travel, and more. Some very long-lasting friendships were made in this shared experience.

What knowledge or new understanding do you hope that a visitor would take away from a trip to Kitt Peak?

For several generations, Kitt Peak has served the needs of U.S. astronomers, particularly scientists who otherwise would not have had access to instruments on which to carry out their science. Speaking personally, my career would have been very different if Kitt Peak had not been available.

Science Highlight

Dark Matter and Dark Energy

The discovery of unseen "dark matter" with many times greater gravitational influence over the Universe than all the stars and galaxies that we can observe is undoubtedly the most unexpected and exciting discovery in the history of Kitt Peak.

In the mid-1970s, Vera Rubin and Kent Ford of the Carnegie Institution initiated a survey of spiral galaxies similar to the Milky Way, aimed at measuring their masses and studying the chemical composition of the gas arrayed in their curving arms. Using a very sensitive instrument called an image-tube spectrograph designed by Ford, the duo obtained spectra of these galaxies arising from ionized plasma analogous to the well-known Orion Nebula in our galaxy. The method they used to measure galactic masses is analogous to the one we use to "weigh" the Sun. It is based on combining the observed velocity of Earth as it circles the Sun in its orbit with its measured distance from the Sun. Any object—like Earth—that remains in orbit around a central star must have a velocity slow enough so that it does not fly off into space, unbound from its parent, yet fast enough that the orbiting object doesn't fall toward the star in a life-stunting "death spiral."

By analogy, the mass of a galaxy can be estimated by measuring the velocities of regions similar to the Orion Nebula as they orbit the center of a galaxy, in combination with the distances of these regions from the galactic center. Rubin and Ford made use of what was then the most powerful spectrograph in the world, fed by the world-class light-gathering power of the 2.1-meter and Mayall 4-meter telescopes on Kitt Peak, to measure the orbital velocities of these nebulae with great precision, while imagery from photographic plates provided a clear measure of their distance from their galactic center.

The masses of galaxies they derived from these measured quantities exceeded, sometimes by a factor of 10 or more, the masses that were estimated from measuring the total brightness of the galaxy. The apparent discrepancy between the galactic masses derived from the orbital velocity of nebulae and the masses derived from observed galactic luminosities led astronomers to propose the mind-bending existence of some sort of "dark matter"—in other words, something that contributes significantly to the gravitational pull on orbiting stars and nebulae in a galaxy but does not produce detectable light. From careful studies of large numbers of galaxies, as well as the motions of galaxies in clusters of galaxies, astronomers now estimate that dark matter comprises 90 percent of the total mass of the Universe; its nature is still a matter of vigorous debate and inquiry among astrophysicists.

More recently, several Kitt Peak telescopes have played a major role in the discovery of an even more bizarre "dark" force—dark energy. Unlike the gravitational force of ordinary matter, the "gravity" of dark energy is repulsive. Some six billion years ago, it began to overwhelm ordinary gravity and push the Universe apart.

The presence of dark energy was first recognized from studies of the relationship between the velocities of galaxies and their distance from Earth. In the 1920s, Edwin Hubble, after whom the well-known earth-orbiting telescope is named, showed that galaxies analogous to the Milky Way appear to be receding from us at velocities proportional to their distance away from Earth. Velocities are measured from spectroscopy of integrated light arising from stars or nebulae in distant galaxies, while distance is measured by using an object whose intrinsic brightness is well known from calibrating observations—a so-called standard candle. By comparing the apparent brightness of stars with known brightness fluctuations (so-called Cepheid variable stars; see page 9) to their known intrinsic brightness, the distance to the object containing the candle can be determined by comparing how bright the object appears to be versus its true calculated brightness, which is proportional to a fraction: 1 over the distance between them squared.

Determining the distance to galaxies located billions of light-years from Earth requires an enormously bright standard candle (far brighter than a Cepheid variable) in order that the dim light from such objects be visible from Earth. Supernovae—exploding stars that ended their lives with masses ten or more times the mass of the Sun—can shine with a brightness 100 million times that of the Sun during the first days and weeks after they explode. Astronomers have studied the properties of supernovae in galaxies of known distance and, by tracking their brightness as a function of time, have found that certain types of supernovae with characteristic brightness-time patterns have nearly identical intrinsic brightnesses at their peak. Teams of astronomers working with ground-based telescopes at Kitt Peak's sister observatory in Chile, Cerro Tololo Inter-American Observatory, and with the Hubble Space Telescope were able to find supernovae in a sample of distant galaxies. Together, these observations helped astronomers to determine their velocities of recession from Earth via spectroscopic measurements and their distances via light curves of supernovae, many of which were observed from the 3.5-meter WIYN Telescope on Kitt Peak.

These distant galaxies departed noticeably from the linear relationship between velocity and distance found for nearby galaxies. Instead, they were found to be moving away from Earth significantly faster than a simple extrapolation of that linear relationship observed nearby. Two different teams came to this surprising conclusion at the same time, helping give confidence to the imposing conclusion that there must be some force that had pushed these galaxies farther apart than the existing mainstream cosmology would allow.

Ironically, Albert Einstein first imagined a repulsive force pervading space in a landmark paper he published in 1917, but he abandoned this notion after Hubble's discovery of the local linear relationship. Far from his "greatest mistake" (as some versions of Einstein's life story suggest that he considered it), Einstein's initial insight seems to be confirmed by the properties of dark energy that we have inferred so far.

The nature and cause of this force are a mystery: one that challenges the imaginations of scientists throughout the world and, in tandem with dark matter, presents a fundamental challenge to both astronomers and particle physicists.

A daytime guided tour of Lowell Observatory's historic Clark Telescope.

Lowell Observatory

Flagstaff, Arizona

A Private Vision Producing Popular Science

Lowell Observatory, a vibrant research institution for nearly a century, has the longest and perhaps most colorful history among observatories of the American Southwest. This private, nonprofit research institution sits on a large mesa named Mars Hill, directly above downtown Flagstaff, Arizona, but even the short drive up is enough to firmly leave the town behind for the peaceful isolation of the observatory complex.

One of the earliest scientific outposts in the Arizona Territory, Lowell Observatory was founded in 1894 by Percival Lowell, a wealthy Bostonian eccentric securely in the top rank of uniquely American supporters of the sciences. Though influenced even now by Percival's lifelong outsider status, the observatory has been the site of many important scientific discoveries, from the fundamental measurements of the recessional velocities of galaxies in 1912–1914, which ultimately inspired Hubble to make the observations that demonstrated that the Universe is expanding, to the culture-tweaking discovery of Pluto. The observatory then entered a relatively quiet period where its staff was dominated by three

Web site
www.lowell.edu
www.lowell.edu/outreach

Phone
(928) 233-3211

Address
1400 West Mars Hill Road
Flagstaff, AZ 86001

aging early heroes, until a one-two punch of new directors in the 1950s vaulted the observatory back to the prominence it once enjoyed.

Today the observatory's twenty-person scientific staff is a world leader in many areas of astronomy and space science, particularly in studies of galaxy evolution and the smaller bodies in our solar system: rocky asteroids and distant icy bodies revolving around the Sun in a region outside Neptune's orbit known as the Kuiper Belt. Scientists at the Lowell Observatory typically use a wide variety of telescopes, ranging from the five primary research telescopes owned by Lowell to an assortment of spacecraft in Earth orbit and beyond.

The observatory is home to a melange of telescopes—four of the largest are located off Mars Hill at a second, darker site east of town called Anderson Mesa (see "For the Public," below, for more information on Anderson Mesa). Mars Hill will always be known first as the home of the historic 24-inch (0.6-meter) Alvan Clark Telescope (named for the famous late-nineteenth- and early-twentieth-century craftsman of large telescope lenses), which remains the showpiece and heart of Lowell Observatory. Flush with his semiprofessional passion for Mars, Percival Lowell had the Clark Telescope installed on Mars Hill in July 1896 for a capital investment of $20,000, establishing what is generally acknowledged as the first permanent major scientific instrument in the Arizona Territory. The telescope dome once floated on salt water and was rotated by pulling on ropes, a distinctly nautical solution to smoothly rotate such a large mass; today, it uses tires from a 1954 Ford pickup truck.

Nearly two decades after it began work, V. M. Slipher used this refracting telescope to discover the first evidence that the Universe is expanding rapidly in all directions, as if the galaxies were the tiny fruit in a gigantic loaf of rising raisin bread.

During the early 1960s, the Clark Telescope was used by legendary planetary scientist Eugene Shoemaker and others to help create the basic drawings for a landmark effort to make comprehensive, detailed maps of the Moon sponsored by the U.S. Air Force. Soon after, Apollo astronauts were able to make some of their first good observations of their historic future destination through the Clark (as well as the McMath-Pierce Solar Telescope at Kitt Peak). The Clark refractor continues to excite the

The Clark Dome houses the Clark Telescope, purchased by Percival Lowell for studies of the solar system in 1896 from Alvan Clark and Sons in Cambridgeport, Massachusetts.

The Lowell Observatory 24-inch Clark Telescope.

public today during the regular stargazing sessions hosted by the Lowell public outreach staff, and it is a frequent backdrop for many astronomy documentaries of all sorts.

Another historic, relatively small instrument at Lowell is a 21-inch (0.5-meter) telescope housed in a 1920s stone building on Mars Hill. It has been used with astounding patience and dedication to precisely monitor the brightness of Uranus and Neptune for more than fifty years, as well as Saturn's moon Titan over Saturn's full three-decade orbital cycle. Originally designed to monitor the total energy output of the Sun, this relatively tiny telescope was also used for historic work that defined the first widely accepted process for categorizing stars according to their colors (as measured through filters that isolated light at near-ultraviolet, blue, and visual wavelengths), and it produced a decades-long series of sky-brightness measurements that helped Flagstaff be named the world's first International Dark-Sky City.

Similarly, the landmark discovery of Pluto was accomplished at Lowell using a 13-inch (0.3-meter) telescope optimized for photography and known as an astrograph. Kansan Clyde Tombaugh used the 13-inch to discover the much-prophesized "Planet X." Later named Pluto at the suggestion of an English schoolgirl, it is now considered a so-called dwarf planet in a controversial redesignation passed by the members of the International Astronomical Union.

Long predicted to exist by Percival Lowell and others such as William Pickering, Pluto was detected using a machine to "blink" or compare the star fields on photographic plates taken on different nights. In early 1930, the patient Tombaugh used his eyes—not a computer!—to spot the relatively large change in position of the tiny speck of reflected sunlight—the expected signature of a candidate planet. From "blinking" images spanning several nights, he was able to see and measure the apparent motion of the relatively nearby Pluto, as compared to the vanishingly small movement of distant background stars, confirming it as a body lying beyond the orbit of Neptune. Both the astrographic telescope and the plates used to find Pluto are on display today in Lowell's Rotunda Library Museum.

Flagstaff continues to have amazing contemporary links with Pluto and its companion moons. Pluto's major moon, Charon, was discovered

The 13-inch Pluto Discovery Telescope is open to visitors for specially arranged tours.

in 1978 by Jim Christy of the staff of the U.S. Naval Observatory in Washington, D.C., using observations made in Flagstaff. More recently, Marc Buie of the Lowell scientific staff was part of a team that used very long exposures taken by the Hubble Space Telescope to discover two tiny new moons of Pluto, since named Nix and Hydra. Will Grundy and other staff members at Lowell are involved in planning the observations of Pluto to be made by NASA's *New Horizons* spacecraft. Already on its way, *New Horizons* will fly past the icy dwarf planet and its expanding coterie of moons in July 2015.

Why does Lowell Observatory exist at all? It is tied inextricably to the Lowell family (who continue serving as its mandated trustees), beginning with Percival. Known as a man of "pure action" from his college days at Harvard University until his death forty years later, Percival had an entirely separate career as an eminently respected scholar of Japan and the Orient (serving as the counselor of the first Korean mission to the United States) before astronomy took full hold of his imagination.

Percival's astronomical obsession appears to have been inspired by several events, starting with a supposed viewing of Donati's comet in 1858 when he was three years old. At age fifteen, he received a 2.25-inch (0.06-meter) brass telescope from his mother (on display today in the Lowell Rotunda Library Museum), which he first used on the roof of the family home in Brookline, Massachusetts.

Twenty-four years later, Lowell used this 30-inch-long instrument to view Gale's comet—a comet discovered in the Southern Hemisphere but first sighted in the Northern Hemisphere by Andrew Douglass in Flagstaff. Douglass had been dispatched by Lowell to survey the rugged Arizona Territory for an appropriate site for a world-class observatory. (He sent word of the comet sighting back to Lowell via coded telegram, early evidence of the secretive passion that characterized Lowell's approach to his astronomical pursuits.)

Lowell was also colored by his contact with the eccentric French astronomer Camille Flammarion, a prolific author and eager popularizer of scientific ideas who was a vocal advocate of the theory of the "plurality of worlds"—the idea that the conditions for extraterrestrial life are common. Whether this was accomplished by theological generosity or by fully agnostic chemistry was a matter of great discussion that continues to echo today. Lowell searched for evidence of life on other worlds, focusing his attention on Mars, where he believed (incorrectly as it turns out) that a web of fine features he observed on the planet's surface was evidence of canals constructed by an advanced civilization. While the ubiquity of life beyond Earth still remains a subject of conjecture, the more than three hundred extrasolar planets detected by astronomers in recent years using several techniques would surely please and embolden the pluralists.

Lowell died on November 12, 1916, in Flagstaff, and he is buried on Mars Hill. His impressive cement tomb overlooks Flagstaff and resembles a large antique globe, evoking both the iconic power of an observatory and the grand dome of stars in the night sky. The funds spent on building the elaborate tomb were resented at the time of its construction by staff astronomers, who would have preferred that Lowell's widow spend the money on new research equipment.

It is inscribed with a much-referenced quotation of Lowell's emphasizing the "bodily isolation . . . and purity" demanded of those brave and lonely men who dare "to see into the beyond." Percival lamented with undisguised pride that such creatures must suffer a hermit–like existence in order to do it, aloof from urban humanity and certain to be disbelieved when returning from the mountaintop with their newfound wisdom.

The future of Lowell Observatory is intimately entwined with a project that, although born of much more modern sensibilities, has again been supported significantly by the passion of one wealthy individual. The $40 million Discovery Channel Telescope (DCT), under final construction forty-five miles southeast of Flagstaff at a site in the Coconino National Forest known as Happy Jack, was born via an interesting marriage between the private observatory and the multimedia company Discovery Communications. In particular, it has benefited from a $6 million donation from John Hendricks and family. Hendricks, the founder and chairman of Discovery Communications, has a personal interest in astronomy and space science and has been a member of the Lowell Observatory Advisory Board for more than a decade.

The DCT, situated at an altitude of 7,760 feet, has a purposefully flexible design that makes it capable both of a very wide field of view (four times the apparent diameter of the full Moon) at its prime focus and aimed primarily for planetary science research, and of a different optical configuration (called Ritchey-Chrétien) that is better suited for research that demands high-resolution imaging or spectroscopy.

The primary mirror for the DCT weighs "only" 6,700 pounds, very light for a such a large mirror—the thin slab of ultra-low-expansion glass is 169 inches (4.3 meters) in diameter but less than 4 inches (100 millimeters) thick. It was cast and fused at Corning's plant in Canton, New York, and polished to an accuracy of a few millionths of an inch by the University of Arizona's College of Optical Sciences.

Construction at the site began in mid-2005, and it is planned that the telescope will be operational in 2010, at which point it would enjoy the honor of being the fifth-largest telescope in the continental United States. A study by the Center for Business Outreach at Northern Arizona University found that the DCT will contribute more than $576 million to the economy of the state of Arizona over its planned fifty-year useful lifetime—a surprising but noteworthy conclusion that bespeaks the importance of astronomy not only to "pure" research but to stimulating other scientific, technical, and economic activity.

Once operational, the DCT will certainly expand the legacy of Lowell Observatory's groundbreaking discoveries of distant bodies in our solar system, and beyond. It should be easy to follow the project's progress: The telescope and the research it enables will be the focus of ongoing educational television programs about astronomy, science, and technology to be aired on Discovery's multimedia networks.

For the Public

Located just one mile west of historic downtown Flagstaff, Arizona, at an elevation of 7,200 feet, Lowell's scenic Mars Hill campus draws more than 70,000 visitors a year from the local community and from Phoenix and California, whose residents look to escape the local summer heat.

The staff of the 6,500-square-foot Steele Visitor Center offers guided thirty-minute tours of the observatory daily on the hour from 10:15 A.M. to 4:15 P.M. (1:15 P.M. to 4:15 P.M. in winter).

The main exhibit hall and science center located in the Steele Visitor Center at Lowell Observatory features exhibits such as "The Galaxies" and "The Cosmos."

The Rotunda Library Museum at Lowell Observatory houses many historic exhibits and a hands-on exhibit for children. The museum is part of a guided daytime tour for visitors and is often open during special events.

The John Vickers McAllister Space Theatre opened in May 2007. Tucked inside the existing visitor center exhibit gallery, the twenty-four-seat theater presents a twenty-minute narrated program during the daytime every hour starting at 10:00 A.M. through 4:00 P.M. (1:00 P.M. to 4:00 P.M. in winter). Lowell outreach staff make an effort to add lively music and timely news bits regarding the latest astronomical discoveries to the program, which is projected on a concave five-foot-diameter screen inside the planetarium-like venue. At night, the theater shows a shorter program to star-party guests designed to give them an introduction to what objects they'll be trying to see in the sky later that evening.

Exhibits in the Rotunda Library Museum were restored in 2006. Subjects include the Lowell family, some of Percival Lowell's sketchbooks and his calculations predicting the orbit of Planet X, the process of naming planet Pluto and a sampling of the letters the observatory received after the discovery of Pluto, current research news, and a children-focused exhibit on the workings of a telescope, with lens and eyepieces to play with.

Joining the Friends of Lowell Observatory support group is a great way to participate in special tours of Lowell Observatory's semiremote dark site at Anderson Mesa, about 15 miles southeast of Flagstaff. It is the home of the 1.8-meter (71-inch) Perkins Telescope, the 1.1-meter (43-inch) Hall Telescope, the National Undergraduate Research Observatory, and the Naval Observatory's Prototype Optical Interferometer. Known as NPOI, the interferometer is a collaboration between Lowell Observatory, the Naval Research Lab, and the U.S. Naval Observatory. The interferometer has been used to measure precise diameters and masses of stars and the often-complex properties of binary or multiple star systems.

The Perkins Telescope was originally located near Columbus, Ohio, where it was the third-largest telescope in the world when dedicated in 1935. It was moved to Lowell's Anderson Mesa site in 1960 under an agreement between Ohio Wesleyan University, Ohio State University, and John S. Hall, who was then the director of Lowell, and has been steadily upgraded since then, most recently in a partnership with Boston University.

For Teachers and Students

Lowell staff, including the astronomers, share a dedication to involving teachers and students in the mission of the observatory. Staff scientist Deidre Hunter is particularly known for her active outreach programs with the Navajo and Hopi Nations.

Private, paid, one-hour daytime programs are available by reservation, including two multimedia presentations and a tour. Topics offered include the planets, comets/asteroids/dinosaurs, life on other worlds, and the expanding Universe.

Private evening programs, including exclusive access to the Clark Telescope, are also available.

A Talk with Robert Millis

Former director, Lowell Observatory

Born and educated in the Midwest, Robert Millis spent a summer at Lowell Observatory four decades ago as a graduate student and—seduced by the natural beauty of the area and the scientific freedom of the institution—he has been part of the staff essentially ever since. From 1989 to mid-2009, he served as director of the observatory. He is also a vocal advocate for the economic benefits of astronomy to the state of Arizona and the value of protecting its world-renowned dark skies.

When did you first encounter Lowell Observatory?

I first went to Lowell in 1965 while a graduate student at the University of Wisconsin, where the director at that time, John S. Hall, invited me to do some joint research with him. I joined the scientific staff in the fall of 1967 and have been there ever since.

Growing up in Illinois and not having traveled much, the western United States was a revelation to me, and Flagstaff in particular—its scenic beauty and cultural diversity. Lowell Observatory was very attractive because it was small, very collegial, and the science staff had tremendous

Robert Millis, former director, Lowell Observatory.

freedom to work on whatever interested them. I just really liked the environment, and so did my wife. We arrived there on something like the fifth day of our honeymoon, so we encountered Lowell at a special time in our lives. I always imagined that I'd end up at a university, teaching, but time passed and it was a satisfying place to be, so thoughts of a university career just kind of faded away.

My wife and I lived on the observatory grounds for roughly thirty years. It was a very desirable place to raise our two kids. It's a very parklike surrounding—the forest goes off for miles in one direction, but you could be in the heart of downtown Flagstaff in just five minutes. It's kind of the best of both worlds. Some of Lowell's major observing facilities are outside of the city at Anderson Mesa, but we spent most of our time on Mars Hill.

I gradually became part of management. Back in those days, the observatory entrusted positions of responsibility only to astronomers.

For several years, I served as secretary/treasurer, supervising the bookkeeper who kept the financial records and managed grants and contracts. The total staff is small now, and it was smaller then. A lot of people tended to wear more than one hat. To a degree, that's still true today.

Do your astronomers get special access to the observatory's telescopes or do you also accept visiting astronomers?

We currently have twenty-one PhD scientific staff. First and foremost, our telescopes exist to serve them. We have always hosted visiting astronomers and will continue to do so, but staff has priority, though some staff members rarely if ever use our facilities. They go to the national observatory near Tucson [Kitt Peak] or other telescopes on Mauna Kea, and so forth.

Have you always been interested in solar system studies? Why do you find it so fascinating?

I came into planetary astronomy through the back door. It was not part of the curriculum at the University of Wisconsin. My thesis was on variable stars, but one thing I learned was how to do precision photometry [measuring tiny variations in starlight brightness]. So when I joined the Lowell staff, I began to look for applications in planetary science where my skills could be applied. I began a long history of studying the smaller bodies in the solar system, on through the satellites of the outer planets, asteroids and comets, ring systems, and Pluto, finally arriving out at the Kuiper Belt in the late 1990s. The unifying thread has been my interest in the smaller objects.

I tell people that I have a short attention span. I like to study things that change, that are dynamic. The moons of the outer planets show changes during their orbits, and there is tremendous variation among them. When I was first studying them, they were just specks of light in a telescope, not the intriguing worlds that the *Voyager* spacecraft showed us.

I was lucky enough to be part of the discovery of the rings of Uranus in 1977, which brought some NASA interest and funding. With Jim Elliot at MIT and others, we had great fun chasing occultations of stars (eclipses caused by an asteroid passing in front of a background star) to directly calculate the sizes of asteroids and help calibrate other observations. We were part of the first team that was able to observe the occultation of a star by Pluto, which enabled the first direct detection of the profile of its atmosphere (basically its density as a function of height).

Then along the way, Mike A'Hearn from the University of Maryland and I started on a long-term study of the group properties of comets, a program which persists today at Lowell under the leadership of Dave Schleicher. We've been able to observe about 120 comets, all on a very standard measuring system designed to determine differences in chemical composition between comets and get a better understanding of how comets respond to the changing level of solar heating as they go around their orbits. It turns out there's a lot less diversity than one might have imagined, but there are a few real outliers [oddballs] and we still don't really have a clear explanation of how they came to be so different in composition.

By the mid-1980s, I was beginning to get discouraged about the prospects for ground-based planetary astronomy. Then, the first Kuiper Belt object was discovered in 1992 and that opened up a whole new frontier for exploration of the bodies populating our solar system beyond the orbit of Neptune, an exploration which we are really just beginning. There are certainly substantial objects out there that have not yet been discovered, but they will be found over time.

We formed a multi-institution team that won access to the Mayall 4-meter telescope on Kitt Peak to do a survey to discover orbiting objects in the Kuiper Belt (known to astronomers now as "KBOs"). This program helped clarify the dynamical history of the Kuiper Belt (how the pull of gravity of the other planets in the solar system rearranged the orbits of KBOs over time) and provided a large sample of KBOs with well-known orbits that could be studied by telescopes on the ground and in space—a sample crucial for understanding the diversity of orbits and chemical composition among these objects, as well as their likely origin. It has been very satisfying, and Lowell astronomers continue to refine orbits so that our successors can study them years, decades, and centuries hence.

What are the roots of the Discovery Channel Telescope project?

The DCT was conceived during the early days of this period. The exploration of the solar system was very much on our minds. The desire to explore the Kuiper Belt and to contribute to the problem of finding near-Earth asteroids [asteroids whose orbits bring them close to Earth] were the primary reasons for the wide-field capabilities of the telescope.

Lowell Observatory celebrated its centennial in 1994, and it was a natural time to do some soul searching. We could have become more of a think tank, but we decided to continue to be an observatory with significance on a national and international scale. To do that, we decided that we needed a larger and more capable telescope.

Discovery Channel joined the project in 2003, and it really gathered steam. We had the resources to go full-speed ahead on building a 4-meter telescope—which, when you're talking about something as challenging as a 4-meter telescope, is not very fast, I have learned.

What lies in the future for astronomical research at Lowell?

There is a transition occurring as my generation steps aside and another one steps up. Current staff and others will use the DCT and other facilities to continue our studies of the outer solar system. We're also developing more and more expertise in searching for extrasolar planets using the transit technique [watching for the light of a star to dim ever so slightly when a planet passes in front of it in our view], led by people like Ted Dunham. Ground-based telescopes will be the key to sifting through the planet candidates found by the NASA Kepler mission. Other people like Deidre Hunter study star formation and evolution. Another project that began before I arrived and continues today is monitoring the activity level of about one hundred Sun-like stars to better understand the wide range in the energy output of such stars and, by implication, the Sun. I'm sure that will be part of what the DCT does.

What does the Discovery Channel bring to the project?

John Hendricks and his team originally brought the wherewithal to really get the project off the launch pad. What they bring now is the ability to shine a global spotlight on the facility, and the engineering, science, and technology that go into building and operating a modern telescope. My understanding is that Discovery will start with a two-hour documentary on the building of the DCT, and then follow up with programs about the research accomplished with the telescope. They are the experts.

What kind of access will the public have to the DCT?

I suspect there will be intermittent tours after we get operating, as soon as 2010. And there will be a live Web presence in the visitor center on Mars Hill, so a lot of public interaction with the DCT will be through television. It's a pretty remote site, about forty-five miles and a good one-hour drive from Flagstaff, so the DCT is designed to be operable from anyplace in the world, and we believe most astronomers will choose to use it remotely.

What experience do you try to present to public visitors to Mars Hill?

We try to provide a genuine interaction with the visitor, some real person-to-person contact. We want people to go away feeling that Lowell Observatory is a unique and valuable institution with a rich history and a bright future—a bit of a "David" among the "Goliaths." We pride ourselves on being a scientific institution with an open-door policy toward the public. We offer them an opportunity to be as much a part of the observatory as they want, from taking a tour all the way up to being a member of the advisory board.

We let as many people as possible look through the Clark refractor, and most people say it is a supremely satisfying experience. It looks like a telescope should, and the building is pretty funky. It is unusual that we are a private institution and not associated with a university or funded directly by the federal or state government. Most of the other private entities like Lowell have fallen by the wayside.

What is your perspective on the history of the observatory and what some might term its eccentricities?

The current staff feels a genuine connection right back to the founder, Percival Lowell. In part, that is because of the governance structure identified in his will. The observatory is guided by a sole trustee who must be a member of the Lowell family. For the last eighty years, the observatory has been guided by a family member. First it was Percival's nephew Roger Lowell Putnam, then Roger's youngest son, Michael, and now Michael's brother, William Lowell Putnam. Their presence, and that of other family members who visit and talk frequently about "Uncle Percy," gives everyone on the staff a pretty good grounding.

Clearly, Percival Lowell was dead wrong about the existence of intelligent life on Mars, but we can point at other things he initiated that bore tremendous scientific fruit. Nobody can dispute the discovery of the redshifts of galaxies took place using Lowell's original 24-inch Clark Telescope, with the spectrograph he commissioned, by a young astronomer, Vesto Slipher, whom Lowell hired. Whether Pluto is a planet or not, it was discovered in a series of events that were set in motion by Percival Lowell. I like to invite people to stand on Mars Hill and look around at his legacy, and compare it to those of his detractors—Lowell, I think, comes out really well. I often say that Percival Lowell was the Carl Sagan of his day.

What has been the nature of the relationship between Lowell Observatory and the city of Flagstaff?

There has been a long and mutually supportive relationship. The community takes a lot of pride in the observatory and appreciates the visitation it attracts. The city and Coconino County came to the aid of the observatory and our neighbors, the U.S. Naval Observatory, in the mid-1980s when our dark skies were threatened by ongoing population growth by establishing some of the first outdoor lighting ordinances. It is a very astronomy-friendly city, with a large number of people with PhD's. In my opinion, Flagstaff is fortunate to have places like Lowell, the U.S. Geological Survey, the U.S. Naval Observatory, and Northern Arizona University to serve as flywheels for the local economy.

Science Highlight

The Outer Bodies of the Solar System

One of the early landmark discoveries made at the Lowell Observatory occurred in 1930, when Clyde Tombaugh announced the discovery of a planet beyond Neptune: Pluto, a small, distant sphere of rock and ice about one-sixth the size of Earth. Over the subsequent decades, scientists working at Lowell have continued searching for other members of our solar system's extended family. Their research has focused on discovering, analyzing, and categorizing the properties of asteroids, comets, and—more recently—a class of distant icy bodies called trans-Neptunian objects located in the so-called Kuiper Belt.

In retrospect, we consider Pluto to represent the first known example of this group. Collectively, this group of small bodies has a great deal to tell us about the origin and evolution of the solar system and the collisions that have shaped Earth's history.

Over the years, Lowell astronomers have been at the forefront of efforts aimed at cataloging and quantifying the characteristics of asteroids: their sizes, shapes, orbital properties, and surface chemical compositions. These bodies are generally small in diameter, 100 meters to 10 kilometers (328 feet to 6 miles), and they are predominantly located in a loose belt orbiting between Mars and Jupiter. Understanding the properties of asteroids not only provides insights into how these small bodies were formed, but can also reveal complex gravitational interactions among planets and smaller bodies that shuffled the distribution of planetary orbits during the formative phases of our solar system.

Over the 4.6-billion-year history of the solar system, a few percent of the asteroid population has been "scattered" via gravitational interactions with larger bodies and thrust into orbits having a wide range of elliptical shapes and inclinations to the plane defined by the orbits of the major planets. Some of these asteroids are now in orbits that cross Earth's orbit, creating the potential of powerful collisions of the sort that gave rise to the Tunguska explosion over Russia in June 1908 and, much earlier (65 million years ago), to the large impact crater below the Gulf of Mexico thought to have initiated the series of events that led to the extinction of the dinosaurs.

For the past decade, Lowell astronomer Ted Bowell has led a research program aimed at searching for the small subset of asteroids that have the potential of approaching close to Earth. To search for asteroids that pose a potential collision hazard, Bowell has developed an automated telescope at Lowell to search for relatively large "near-Earth asteroids" and determine their orbital trajectories with high precision. Surveys like Bowell's point the way toward more sensitive imaging surveys capable of detecting objects down to a diameter of 150 meters (492 feet) or less; a survey called Pan-STARRS is under way in Hawaii, and a powerful all-sky telescope called the Large Synoptic Survey Telescope is being planned for a site in Chile.

Since the discovery of Pluto by Tombaugh, astronomers had speculated that it was not the only object located beyond the orbit of Neptune. In the early 1950s, pioneering planetary astronomer Gerard Kuiper suggested the existence of an extensive family of objects in these outermost reaches of the solar system. In 1992, his speculation was confirmed by astronomers David Jewitt and Jane Luu, who discovered the first evidence of these bodies in what was soon called the Kuiper Belt.

Subsequent observations have shown that the Kuiper Belt comprises perhaps 100,000 bodies of sizes above 100 kilometers (62 miles) in diameter, with the largest bodies approaching or slightly exceeding the size of Pluto (about 1,000 kilometers, or 621 miles, in diameter). This estimate is based on the statistical distribution of sizes among currently observable objects orbiting the Sun in the Kuiper Belt region. The Kuiper Belt undoubtedly contains a huge number of objects much smaller than 100 kilometers in size; such objects are faint and difficult to observe, so securing an accurate estimate is challenging.

Echoing an original conjecture of Gerard Kuiper's, astronomers now believe that both the large and small members of the Kuiper Belt family were formed in the outer parts of the primordial disk of gas and dust that gave rise to the planets (Mercury through Neptune), which formed in the inner regions of the disk.

The smaller objects in the Kuiper Belt are currently thought to represent the source of comets (such as Halley's comet), whose orbits periodically carry them into the inner solar system. Numerical simulations of the gravitational interactions between the outer planets and Kuiper Belt objects (KBOs) confirm that this scenario can occur by altering (via gravitational interactions with other, larger bodies in the solar system) some fraction of their orbits from their nearly initially circular form to highly elliptical paths that plunge into the inner solar system. Like their rocky cousins in the asteroid belt, some of these comets must have—over the history of the solar system—collided with Earth, possibly bringing the bulk of both the water and the organic material that led to the development of life on our planet.

The distribution of KBO orbits is also beginning to shed light on the evolution of planetary orbits over the history of the solar system. In part as a result of careful analysis of the statistical properties of KBO orbits, astronomers are beginning to entertain seriously the remarkable hypothesis that the planets Uranus and Neptune may well have formed between the orbits of Jupiter and Saturn and migrated outward over time. The agent driving their migration is thought to be the gravitational pulls of Jupiter and Saturn, the two most massive planets in our solar system. As Uranus and Neptune migrated outward, the gravitational force exerted by these two planets affected the orbits of Kuiper Belt objects, sending some of them outward beyond the orbit of Neptune and others inward toward Mars, Earth, Venus, and Mercury.

Some theorists speculate that the "epoch of maximum bombardment" some 300 million to 500 million years after the birth of the solar system, which gave the Moon and Earth most of their craters, was triggered by this massive redirection of the orbits of KBOs.

Already, a growing handful of Pluto-sized bodies with evocative names such as Sedna and Makemake have been found and mapped. Within a decade or two, we may find that our solar system is home to dozens more bodies of Pluto's size or larger, rendering the comparatively lonely "Nine Planets" a quaint relic of the twentieth century.

The converted MMT, featuring a monolithic 6.5-meter primary mirror in place of multiple smaller mirrors.

Fred Lawrence Whipple Observatory

Amado, Arizona

An Iconic Founder and a Giant Telescope of Two Vintages

The location of Whipple Observatory on Mount Hopkins retains far more of its original isolation from the everyday world than most of the sites in this book, more than four decades after it was first scouted. Located in the rugged Santa Rita Mountains, thirty-five miles south of Tucson near Amado, Arizona, the observatory was the brainchild of legendary astronomer Fred Lawrence Whipple of the Smithsonian Institution and Harvard College. The Smithsonian and the University of Arizona jointly operate the largest telescope on the mountain, known as the MMT (for "Multiple Mirror Telescope," a name derived from the telescope's original unique configuration comprising multiple smaller reflecting mirrors that bring light to a common focus).

The Whipple Observatory Visitor Center is an easy drive south of Tucson just past Green Valley, with the final eight miles ending

Web site
www.cfa.harvard.edu/facilities/flwo/visit_center.html

Phone
(520) 670-5707

Address
670 Mount Hopkins Road
Amado, AZ 85645

Observatory Ridge at Mount Hopkins features several 1-meter-class telescopes, a 10-meter optical reflector, and the HATNet automated small telescopes (foreground).

just within the edge of the Coronado National Forest. In addition to the 6.5-meter (21.3-foot) MMT, Mount Hopkins is home to the 1.5-meter (60-inch) Tillinghast Telescope as well as a 1.2-meter (48-inch) and an automated 1.3-meter (51-inch) telescope dedicated to observing the Universe at infrared wavelengths (1 to 20 microns or up to 40 times longer than the wavelengths of light that the human eye can sense).

Other unique facilities include the now-retired Infrared-Optical Telescope Array (IOTA), a pathfinder in the emerging science of astronomical interferometry, and the Hungarian Automated Telescope, part of a group of six automated small telescopes here and two on Hawaii's Mauna Kea that observe variable stars and search for extrasolar planets. In late 2007, a group of 16-inch (0.4-meter) telescopes was installed in the former Mount Hopkins Observatory satellite-tracking station to conduct a search for Earth-size planets around nearby M-class stars (stars much cooler and less massive than the Sun). Watching patiently for the precise dip in light from a star that signals the passage of a planet (or planets) in front its disk, the "M-Earth" survey should be able to detect habitable planets down to twice the radius of Earth.

The roots of Whipple Observatory reach back to 1966 and the height of the "Space Race." Fred Whipple, the director of the Smithsonian Astrophysical Observatory, based in Cambridge on the grounds of the Harvard College Observatory, decided to relocate one of his group's satellite-tracking stations from New Mexico to Arizona. Though the move was rationalized as an escape from growing light pollution near the city limits of Las Cruces, it was based largely on a desire by Whipple (who was primarily a meteorite scientist at that time) and his colleagues to break into the rapidly emerging field of astrophysical research. The rugged look of the original group in New Mexico was captured on an iconic cover of *Life* magazine from February 17, 1958, featuring a metallic-hued color image of two staff technicians gazing skyward alongside their huge, conically shaped telescopic camera as they prepare it to scan the sky for newly disconcerting appearances of Earth-orbiting satellites.

Whipple directed a rapid search of new sites in New Mexico, Arizona, and Baja California with the help of close friend and associate Nicholas Mayall (for whom the 4-meter telescope on Kitt Peak is named). The Smithsonian settled on his preferred site in the Coronado National Forest near Amado, Arizona, south of Tucson, on 8,555-foot Mount Hopkins. Whipple personally checked out the air turbulence near the site as a passenger in a light plane piloted by fellow astronomer Gerard Kuiper.

Soon after being chosen as the site, the mountaintop was subject to a five-foot snowstorm in December 1967, the worst such event since 1914. But this powdery blast did not faze Charles A. "Chuck" Tougas, the initial Mount Hopkins Observatory site manager and "strong man of the 'frontier era' of the observatory," according to Whipple. A colorful and boundlessly energetic character who was one of the two men pictured on the *Life* cover, Tougas cleared the snow driving a road grader borrowed from the expert road crew that blasted and bulldozed the path to the snowy summit. Renowned for his creative local trades of material and mechanical services that enabled Whipple to be built on a relatively shoestring budget, Tougas made it clear from his opening memos on new staff hiring that such "positive traits" as enthusiasm, willingness to accept challenging situations, and personal compatibility would be required by new staff. "Preferably more of these traits should be present than in a candidate for most of the other existing [Smithsonian] stations," he wrote with confident flair.

The earliest telescopes on the mountain were a large, segmented-mirror reflector dish to look at high-energy gamma rays and a 12-inch (0.3-meter) optical telescope. After an intensive study of the local environmental properties of the five distinctive humps (more pleasantly renamed "knolls" in later detailed survey documents) at the end of its extra-steep and rugged access road, a 60-inch (1.5-meter) telescope was built on "Knoll #1" at a working altitude of 7,600 feet.

Mount Hopkins Observatory was dedicated on October 23, 1968, by dignitaries including U.S. Representative Morris Udall of Arizona. The original 12-inch telescope site was later used as the location for a 24-inch and then a 48-inch telescope.

From the beginning, Whipple Observatory's science targets included sources of high-energy gamma rays, not just visible or infrared light, thus carving out a special niche with facilities like the first Imaging Atmospheric Cherenkov Telescope (which searches for faint transient flashes of light when gamma rays hit the top of Earth's atmosphere). This specialty continues today with a new world-class, gamma-ray-seeking array of four telescopes called VERITAS (Very Energetic Radiation Imaging Telescope Array System), now operating near the visitor center.

In September 1967, one of the defining characters of Whipple Observatory, Irishman Trevor Weekes, arrived from Smithsonian headquarters in Cambridge, Massachusetts, and quickly started demonstrating the promise of the site for observing high-energy gamma rays with a 10-meter (394-inch) telescope called the Large Optical Reflector (LOR). Capturing gamma rays does not require a mirrored surface as smooth as those for visible light, so the LOR was augmented occasionally with World War II-era military-surplus searchlight dishes retrofitted as telescopes.

Weekes later served as director of the observatory until 1976. He was awarded the prestigious Rossi Prize in 1997 for his discoveries of the first galactic (inside the Milky Way) and extragalactic (outside the Milky Way) sources of very-high-energy gamma rays, and he continues

The four 12-meter reflectors of the Very Energetic Radiation Imaging Telescope Array System (VERITAS) positioned around the Whipple Observatory Visitor Center and office complex.

to work at the scientific forefront today in a leadership role on science from VERITAS.

In May 1969, astronomer George Rieke received the first Harvard PhD based on data primarily from Mount Hopkins, again on high-energy gamma rays. Today, Rieke is half of a world-renowned scientific team with his wife, Marcia. Based at the University of Arizona's Steward Observatory, they lead large teams that produce world-class instruments for space-based NASA observatories like the infrared-optimized Spitzer Space Telescope and the under-construction successor to the Hubble Space Telescope, called the James Webb Space Telescope.

A series of public lectures in the late 1960s, consistent with the general outreach spirit of the Smithsonian, were initiated by Weekes at the Amado School House, with priceless "signs-of-the-time" titles like "The Sixties: Astronomy Turns On" by Rieke. Smallish crowds attended these talks, but, even by this time, the lights of neighboring towns had begun to infringe upon the site, making such nascent efforts at local public outreach even more beneficial.

In some cases—for example, the glare from a neighboring dog racing track—the observatory staff even pitched in to help install better, more shielded fixtures. Weekes soon met with the mayor of Tucson, along with the directors of Kitt Peak and Steward Observatory, leading eventually to landmark legislation on control of urban light pollution. Longtime Whipple public information officer Dan Brocious remains at the vanguard of ongoing policy work to retain the effectiveness of these regulations as the region around the observatory becomes increasingly developed and subject to possible incursions of bright lights from U.S. Border Patrol checkpoints.

The 1.5-meter telescope at Mount Hopkins was finally dedicated in September 1970, the same month that 82-mile-per-hour winds were recorded at the site. The telescope was named after the late Carlton W. Tillinghast Jr., a legendary early administrator at the Smithsonian Astrophysical Observatory. The 1.5-meter's dome is connected to the 1.2-meter's dome by a somewhat odd shared support building. The 1.5-meter Tillinghast was used most famously in a landmark long-term effort by John Huchra and Margaret Gellar to map thousands of galaxies across a significant slice of the sky that led to the profound realization that the Universe is populated by a web-like distribution of galaxies, where large voids occupy vast regions separating clusters and superclusters of galaxies.

For a period in the 1970s, the primary focus of the Smithsonian Observatory headquarters shifted toward a new specialty in the quickly emerging field of x-ray astronomy from space satellites. But out in the western desert far from the Harvard campus, astronomer Aden Meinel worked on his personal contacts in the world of large telescope optics to obtain six 1.8-meter (71-inch) mirrors originally intended for a space station to be staffed by air force astronauts functioning as orbital spies in the sky. Meinel's goal was to bring the light from each of the six mirrors to a common focus and, by so doing, achieve the light-gathering power of a 4.5-meter (177-inch) telescope.

Meinel (influential in the founding of both Kitt Peak and the Optical Sciences Center at the University of Arizona) and Fred Whipple were the central "parents" of this Multiple Mirror Telescope (MMT). But many others, such as Frank Low (a pioneer of infrared astronomy at the University of Arizona) and Nat Carleton (originator of the idea of the rotating building), played key roles in shepherding this unique telescope to completion.

The modern era at Mount Hopkins began symbolically in March 1976, when a time capsule was placed in the pier of the new MMT to formally mark the start of construction. The structure of the MMT was innovative in every sense. The boxy exterior is four stories tall, yet the entire massive 450-ton building (as opposed to just the telescope structure) rotates on four wheels during astronomical observations. Located on a 8,555-foot knoll 900 feet above the rest of the mountain's telescopes on Observatory Ridge, the building contains a generous 7,500 square feet of usable space around the 90-ton telescope, whose compact alt-azimuth design (similar to the WIYN Telescope at Kitt Peak; see page 36) leaves it squat enough to fit economically inside the exterior box.

Light gathered from the six mirrors was combined at a common focus to mimic the power of a single mirror of size on the order of 4.5 meters in diameter, achieving this goal at much lower cost than by fabricating a monolithic mirror this large with the mirror-making technology of the time. This combined power made the MMT effectively the second-largest telescope in the world when it was dedicated on a clear day in May 1979 by a crowd including Apollo 11 command module pilot and Smithsonian National Air & Space Museum director Michael Collins. The power of the MMT was surpassed only by the behemoth 200-inch Hale Telescope at Palomar Observatory and a 6-meter (19.6-foot) telescope in southern Russia that has never reached its full scientific potential.

The original MMT produced data that contributed to more than a thousand scientific papers, including a major survey of quasars (the incredibly bright and energetic cores of distant galaxies, which are presumed to be powered by massive black holes; see page 9), detailed studies of the remnants of supernovae, and the discovery (by George and Marcia Rieke) of water ice on Pluto's moon Charon.

However, by the late 1980s, major advances had been made in the "spin-casting" approach (developed by world-renowned astronomer Roger Angel) to making and shaping large astronomical mirrors, particularly in the spinning oven of the University of Arizona's Steward Observatory Mirror Laboratory located under the east stand of the football stadium. This leap forward in the economics and quality of large mirrors made it scientifically attractive to replace the six smaller mirrors comprising the original MMT with a single monolithic mirror. A $20 million conversion project in partnership with the University of Arizona resulted in a 6.5-meter primary mirror that more than doubled the MMT's total light-collecting area and expanded its field of view by fifteen times.

The new primary mirror was installed in March 1999, and a second-generation MMT for the new century was dedicated on May 20, 2000. An internal staff contest to rename the telescope failed to arrive at a satisfactory replacement, so the unlyrical "MMT" remains as its name. One of the original 1.8-meter mirrors in the boxy telescope was sent to Kitt Peak, where it became the basis for one of the two Spacewatch telescopes that conduct some of the most productive surveys in operation today for detecting and characterizing the orbits of near-Earth objects such as asteroids that could strike the Earth (again!) someday. The other five primary mirrors remain stored in a warehouse at the Amado base camp.

One signature instrument today at the MMT is Megacam, a 340-million-pixel digital camera based on a mosaic of 36 individual CCDs (charge-coupled devices). A single image from Megacam covers one-half degree in the sky, roughly equal to the apparent size of the full Moon. The modernized MMT has been used to observe some of the most distant galaxies known, at a cosmological "redshift" of more than six (which equates to a time that is less than a billion years after the Big Bang). On the other end of the cosmological scale, MMT data have helped astronomers find one of the lowest-mass white dwarf stars ever seen (at about one-fifth the mass of the Sun). A typical white dwarf star is comparable in diameter to planet Saturn but has a surface temperature exceeding 17,000 degrees Fahrenheit (almost three times that of the Sun). Other MMT observations detected two runaway stars ejected outward from our Milky Way galaxy at more than a million miles per hour.

Whipple Observatory has also been a pioneer in the increasingly important area of astronomy known as interferometry, in which the light gathered by two or more small telescopes is combined to produce measurements with a resolution (sharpness) equivalent to a telescope with a diameter as large as the largest distance between the much smaller mirrors comprising the elements of the interferometer.

The Infrared-Optical Telescope Array (IOTA) was operated jointly on Mount Hopkins from 1993 to 2006 by the Smithsonian Astrophysical Observatory, Harvard University, the University of Massachusetts, the University of Wyoming, and the Massachusetts Institute of Technology's

Lincoln Laboratory. The initial two-telescope interferometer was expanded in 2000 with the addition of a third 0.45-meter (18-inch) telescope, creating the first infrared-optical interferometer trio in the world.

While radio astronomers routinely use arrays to simulate much larger telescopes, the relatively long wavelengths of radio waves—measured in centimeters or even meters—make it much easier to detect fractional wavelength differences between the arrival times of light at the separated telescopes, which is crucial to combining the light from the individual telescopes and reconstructing an "image." Attempting interferometry in the near infrared is much harder because the wavelengths are 100,000 times smaller then the length of radio waves. But IOTA and other arrays, such as the Naval Prototype Optical Interferometer (NPOI) at Lowell Observatory's Anderson Mesa (see page 59), are giving today's astronomers their first-ever accurate views of the shapes of distant stars (which are often oblong due to high rotation speeds) and, amazingly, of activity on their surfaces, such as sunspots, flares, and periodic stellar pulsations.

Fred Whipple himself did not make any particularly historic observations at Mount Hopkins, but his renegade spirit and perennially bolatied visage define the spirit of the place even today. Known best for the "dirty snowball" theory of comets in the 1940s that proved prescient when such ices were later discovered, Whipple died in September 2004, still active as NASA's oldest "principal investigator," bicycling to work and carrying out imaginative scientific studies almost to the day of his passing at age ninety-seven.

"Astronomy is my avocation as well as my vocation," he told the crowd at the ceremony on May 7, 1982, that renamed the observatory in his honor. Beyond his comet studies, Whipple's pioneering work with data from the early satellite-tracking cameras established new benchmarks in our understanding of variations in Earth's gravitational field.

Whipple is one of the key members of the remarkable pantheon of leaders in the early history of astronomy in Arizona, along with Kitt Peak–related colleagues such as Nicholas Mayall and Aden Meinel and the University of Arizona's Gerard Kuiper and Bart Bok. Though these giants are gone, Whipple Observatory today remains as his legacy—a relatively modest outpost that routinely does big things.

For the Public

Whipple Observatory takes advantage of its location at the boundary of the Coronado National Forest to team with the U.S. Forest Service to present an attractive visitor center that contains an instructive blend of information about astronomy and astrophysics, natural science, and cultural history. In particular, a display on the decades of Smithsonian-sponsored natural science research in southern Arizona during the past hundred years shows that the local roots of the famous institution run deep.

Exhibits include models of the original multiple-mirror version and converted 6.5-meter MMT, a three-dimensional model of galaxy distribution in the Universe, a touchable topographical map of the Santa Rita Mountains, and explanations of the challenging concept of carrying out very-high-energy astronomy by using optical telescopes to detect the "flashes" produced as gamma rays interact with atoms and molecules high in Earth's atmosphere.

A natural history exhibit examines nocturnal animals and features a large color transparency of the night sky over southern Arizona. All exhibits and public areas are wheelchair accessible, and major exhibit titles have been translated into Spanish. A full-text, bilingual guide to selected exhibits is available, and presentation videos inside the visitor center are open-captioned for those with hearing impairments.

Completed in 1991, the visitor center is open from 8:30 A.M. to 4:30 P.M. Monday through Friday, except major holidays. The complex includes an outdoor patio with a Native American petroglyph discovered on-site during construction, interpretative signage describing desert flora, and dramatic views of the surrounding Santa Rita Mountains.

Two spotting devices, a 20-power telescope with an individual adjustable focus and a set of wide-field binoculars with automatic focusing, are installed on the outdoor patio of the visitor center to provide a close-up view of the MMT as it sits atop the distant summit of Mount Hopkins. It is also possible to see the domes of Kitt Peak National Observatory,

Whipple Observatory scientists participate in a variety of public outreach events such as Family Science Nights in the Santa Cruz Valley.

located about fifty miles to the west. The binoculars are wheelchair accessible, which makes them accessible to young children as well.

A trailhead, restrooms, and picnic area developed by the Forest Service and located just outside the main gate are open twenty-four hours a day. There are benches, grills, and a hardened wheelchair-accessible path that leads to vantage points overlooking Montosa Wash, a deep drainage running parallel to the site. A kiosk at the trailhead provides information about camping and hiking.

Special public star parties featuring lectures and telescope viewing are held quarterly at the visitor center on Saturdays, starting late in the afternoon. Amateur astronomers are invited to bring their telescopes to the Astronomy Vista, a special observing site with concrete pads and benches that was installed in 1984 along a knoll at an elevation of

5,000 feet, approximately 1.2 miles east of the visitor center on a paved road. Here, within sight of the MMT, amateurs may take advantage of the same clear, dark skies so important to professional astronomers. (Access to telescope pads requires climbing a short, but somewhat steep, unpaved trail.)

Reserved-seat bus tours, originating at the visitor center, are conducted on Mondays, Wednesdays, and Fridays from mid-March through November. Reservations are required and can be made in advance by calling the visitor center. Reservations are on a first-come, first-served basis up to a thirty-visitor maximum. Tickets are purchased on the day of the tour. Tour participants should dress warmly, bring lunches, and be prepared for moderate exertion at high altitude.

For Teachers and Students

There is no charge for Whipple Observatory tours by schools, youth educational programs, and Scout groups. Call the visitor center to make special arrangements. The observatory also offers programs for local schools.

A Talk with Frederick H. Chaffee

Former director, Mount Hopkins Observatory

Frederick Chaffee was fresh from earning a PhD at the University of Arizona (which included a career-changing exposure to the challenges of operating telescopes) when he arrived at the Smithsonian Astrophysical Observatory (SAO) in Cambridge, Mass., in 1968. The young postdoc soon found himself deeply involved in the transition of the Mount Hopkins Observatory from a remote Smithsonian outpost in the dry desert south of Tucson to a fully staffed astronomical observatory capable of supporting a wide variety of unique facilities. Chaffee was the first director of the overall observatory; he resigned from this

Fred Chaffee, former director, Mount Hopkins Observatory.

post in 1984 (at which time its name had been changed to Whipple Observatory) to take on the even greater challenge of directing the observatory's largest and most ambitious telescope, the MMT. He led the operations of the MMT for twelve years before leaving to direct the world-renowned Keck Observatory on Mauna Kea, Hawaii. In both cases, the basic challenge was to take a very large, newly built research telescope and complete its transition from what was fundamentally an engineering project to a full-blown scientific machine.

What is your perspective on the roots of Whipple Observatory?

At the time of the launch of *Sputnik* [1957], there was no organization charged with tracking artificial satellites. Fred Whipple had anticipated this would be something important to do, and he was ready to jump in. This task needed a location in the Southwest, and Whipple personally selected Mount Hopkins because of its proximity to the other observatories in the area.

When I went to SAO, Steve Strom was there, and he had a very active group in the study of stellar atmospheres, which was then a major scientific topic. Steve had been lobbying hard with Fred Whipple to build a big optical telescope for stellar spectroscopy somewhere in the southwestern United States, and Mount Hopkins was an obvious place to go since SAO already has a presence there. Strom was really a key person to make it happen—he was younger than most people on the staff, he had a ton of energy, and a bevy of sharp graduate students around him. He really pushed hard on Whipple.

I was the new postdoc on the block who knew a lot about telescopes, and the SAO did not have a lot of people at that time who did—the staff was mostly theorists and space scientists. I readily volunteered to go out to Arizona and help set things up there starting in 1969. We started with a 24-inch telescope with a spectral scanner, and then the 60-inch telescope [now the 1.5-meter] came along and really put Mount Hopkins on the map.

Who were some of the key figures in the early days?

Chuck Tougas was one of Whipple's golden boys, along with J. T. Williams, and they were absolutely key players in the history of the observatory—really bright, young engineers who worked by the seat of their pants. Whipple had assembled them as part of a group to build the satellite-tracking network. They were natural leaders, and despite their lack of formal training, they quickly rose to the top. Tougas was an extraordinarily gifted and energetic leader, sort of like a supply sergeant, who could make a gallon of lemonade out of lemon seeds. It was really their can-do energy that helped build Mount Hopkins Observatory—these are the kind of people that you never forget.

What are the roots of Whipple Observatory's specialization in the study of high-energy gamma rays?

Shortly after the first optical telescopes were built, Giovanni Fazio, who has gone on to be mainly known as an infrared astronomer, was interested in high-energy astronomy. He hired a young postdoc named Trevor Weekes to set up the 10-meter Cherenkov gamma-ray detector, and that pretty much has been operating there ever since.

It was such an interesting, eclectic mix. The satellite-tracking program was more like defense work in some ways, then you had the large gamma-ray antennas at the high-energy end, and then the steady growth of the optical telescopes.

I was the resident Smithsonian person in Tucson for many years, so I became the first director of Mount Hopkins Observatory in the late 1970s. Until it grew up, it had been treated as more like a field station, with no scientists or administration there. Already, the Smithsonian and the University of Arizona were talking about doing something big together. Aden Meinel was the director of Steward Observatory then, and the MMT was really his brainchild.

I got into the loop fairly quickly as the local SAO representative, but I was running Mount Hopkins as a fully functioning observatory and was reluctant to get drawn into another building project. Jacques Beckers was the first MMT director, and then he moved on to other things—I succeeded him in 1984, until I left to go to Keck. First light of the MMT was in 1979, and a lot of the bugs had been worked out. Jacques was the perfect kind of guy for the transition from a collection of parts into a working telescope—he was very gifted at that. My strength was enabling good science to come out. It was a natural transition.

In retrospect, how successful was the six-mirror design of the MMT?

The MMT worked very well. It was ahead of its time. It was built on the faith that the laser technology to sense where the mirrors were and feed that information back to adjust the secondary mirrors would come along to get the whole system to work. The current technology wasn't really mature enough. An awful lot of energy was expended. Eventually, digital image processors came along that allowed us to point the telescope at a new part of the sky, see six images, rapidly measure their positions, and then send commands to the secondary mirrors to rapidly align them.

Interferometrically phasing the six mirrors [to bring out very fine detail in one object in a small field of view] turned out to be harder. It was eventually done, but the scientific demand wasn't there. For 90 percent of the optical-infrared science being requested, having the mirrors phased wasn't necessary—you just needed the images coaligned so that light from all of them could be collected accurately. During my early years, I emphasized that approach. As a result, in the engineering community, there was a sense that the MMT never fulfilled its technical goals. But that was a conscious decision that I made. If you just "stack" the images, you get a resolution equivalent to an individual 1.8-meter mirror, but a "light grasp" equal to a much larger mirror. We were in essence a 4.5-meter light bucket, but the images were very sharp. My sense was getting a lot of good science out fast was more important, and that was my goal.

How did the transformation of the MMT to one large monolithic mirror come about?

In the early 1980s, just after I had become director, Roger Angel was starting his mirror-casting effort. University of Arizona astronomer Nick Wolff showed a cartoon at a meeting of just how big a telescope with one mirror could fit into the small MMT building—the small size of the building being one of the many really revolutionary things about the MMT. It was also the first big, working alt-azimuth telescope in the United States, and the building rotates, so it was as compact as possible, which has lots of benefits, including keeping the interior atmospheric turbulence (that can degrade the images) as low as possible.

We did not have the funding to start over, but it did seem that we could probably modify the telescope to do just what Nick showed—put one giant mirror in there and more than double the light-gathering power of the telescope. Irwin Shapiro of the Smithsonian and Peter Strittmatter of the University of Arizona decided to do something together.

Everything took much longer than anticipated. It happened much too slowly, in my opinion. Time does march on. The target completion date for the transition to the 6.5-meter was something like 1991 or 1992, which would have made it the largest telescope in the world for some period of time. The MMT could have really reigned supreme. The Keck effort accelerated much faster than anyone anticipated.

By 1996, I was really quite discouraged, and I felt the MMT had really lost its opportunity to change the scientific landscape. When the opportunity came along to direct the Keck Observatory, I had lost my enthusiasm for the 6.5-meter and moved on. The MMT is a very good telescope, but it did not have the impact that it could have had.

What are the main scientific achievements of the MMT that stick out in your mind?

There were two. We did a big survey of quasars, at the time the biggest that had ever been done. There were six papers published in the early 1990s that really expanded the number of these objects that were known. Today we know that quasars, or QSOs [see pages 21–25], are distant energetic galaxies with massive black holes at their core. A group in Cambridge, U.K., identified candidates and we followed them up with the MMT. This was done by Craig Foltz (who became my successor as MMT director), Paul Hewitt, and myself. It was a very satisfying project, and it made a major contribution to quasar spectroscopy.

The individual discovery that was most exciting was the detection of the first gravitational lens. The spectroscopy that was performed to cinch it was done within a week of when the MMT first went online. University of Arizona astronomer Ray Weyman and others had used Kitt Peak telescopes to examine a number of interesting radio quasars, and had stumbled on something that looked extremely unusual. But the Kitt Peak telescopes did not really have enough light-gathering power to tell what they were.

Ray had been a champion of the MMT. He came to us and said this would be really a cool thing to do with it early. I ended up writing a cover story for *Scientific American* on the discovery. It was a real breakthrough in what has become a major subdivision of astronomy.

A gravitational lens is an intergalactic mirage. A foreground object with massive gravity (such as a galaxy or cluster of galaxies) distorts the space between us and a more distant object, making a kind of lens that acts like a funhouse mirror. This lens can do all sorts of strange things to a point image of the more distant object. When you take an image in the direction of such a lens, you can get multiple images of the distant object.

In this case, the first gravitational lens we discovered was a binary, and you could take the spectrum of light from each part, lay one over the other, and compare it pixel by pixel—they were identical, and that was the key signature of a gravitational lens as had been predicted since the 1930s. We could see the data on the computer screen. We knew what it meant, and I literally went out into the dome and screamed, because I was about to explode. It was one of those eureka moments that make a career in astronomy so rewarding.

What do you hope that a visitor will take away from a visit to Whipple Observatory?

This observatory represents a very interesting scientific progression, from a field station studying artificial satellites, to a very good small 1.5-meter telescope that did the premier survey that first showed the lacy structure of the Universe, to the MMT. The history of Whipple Observatory shows that creative people using a small telescope can revolutionize our understanding of cosmology, and that's an amazing thing. The MMT was a technologically revolutionary telescope, which was the first attempt to use a mosaic of telescopes to make a big one. You would be visiting a place that, in the 1980s and 1990s, really revolutionized astronomy. This sort of era of "cowboy astronomy" has passed, but it was an exciting and ennobling thing to be involved in.

Science Highlight

The Large-Scale Structure of the Universe

For much of the twentieth century, astronomers believed that galaxies were distributed more or less uniformly throughout the Universe. The thinking went that if an observer were able to examine spherical shells centered on the Milky Way, the number of galaxies per unit volume contained within each concentric shell would be approximately the same. To be more specific, although each shell might contain small (tens) or large (hundreds to thousands) clusterings of galaxies, on average the number of galaxies, and the distributions of their shapes and intrinsic brightnesses, would be more or less the same.

This belief was reinforced by decades of careful examination of deep photographs of the sky taken with specially designed wide-field telescopes, such as the Palomar Sky Survey carried out with the 48-inch (1.2-meter) Oschin Telescope in southern California (see pages 11–12). Counts of galaxy types and brightnesses, as well as the number of clusters and groups appearing on a given photographic plate from the Palomar survey, generally looked much the same as on any other plate. Of course, photographic images are a two-dimensional projection of the three-dimensional distribution of galaxies. Without a way of measuring the distance to each galaxy, astronomers could not be certain that the appearance of uniformity represented reality.

The growing public availability of sensitive charge-coupled devices (CCDs) in the late 1970s and early 1980s—detectors similar to those commonly used to record images in digital and cell phone cameras today—made it possible for astronomers to contemplate making measurements that would enable a fully accurate three-dimensional map depicting the distribution of galaxies in space.

Marc Davis, Herb Gursky, John Huchra, and Margaret Geller of the Harvard-Smithsonian Center for Astrophysics (CfA) foresaw the potential of CCDs to help chart the true geography of the Universe. Their approach was to mate a CCD detector to an astronomical spectrograph and exploit its ability to efficiently record incoming photons collected by the Mount Hopkins 1.5-meter telescope, thus recording the spectra of large numbers of galaxies and, by so doing, measuring their distances.

Their technique rested on the understanding, established by Slipher, Hubble, and others working in the early twentieth century, that galaxy spectra contain absorption lines produced by the combined light arising from the millions or billions of stars that comprise a galaxy. These lines arise from well-known, cosmically abundant elements such as hydrogen, magnesium, sodium, and calcium. By measuring the wavelengths of these lines and comparing them to their measured wavelengths in terrestrial laboratories, it is possible to calculate the velocity at which a galaxy is approaching or receding from Earth.

Galactic spectral features that appear more "blue," with shorter wavelengths as compared to their terrestrial laboratory counterparts, would be approaching Earth. Those that appear more "red," with longer wavelengths than expected, are receding from us. Edwin Hubble established this basic approach in the 1920s, demonstrating that distant galaxies are receding from Earth as shown by their "redshifted" spectral lines.

By locating bright stars in other galaxies whose intrinsic brightness could be established by observing analogs within our own Milky Way—so-called standard candles—Hubble, Alan Sandage, and, later, teams of other astronomers were able to calculate the distances of other galaxies with great relative precision. Their approach rested on the fact that the measured brightness of a light source dimmed in proportion to the square of the distance of the source from the observer. For example, a standard candle of known brightness in a galaxy located at a distance of 1 million light-years would appear 100 times brighter than the same standard candle located in a galaxy 10 million light-years away from Earth.

A plot of distances obtained from measurements of the apparent brightness of standard-candle stars of known intrinsic brightness versus their redshifts revealed a linear relationship between redshift (recession velocity) and distance: Thus was born the Hubble relationship.

The CfA team using the Mount Hopkins 60-inch telescope exploited the Hubble relationship and the power of CCD detectors to chart the distances of tens of thousands of galaxies out to a distance of several hundred million light-years, covering a vast "pie slice" of the sky. What they found was truly startling and unexpected: a Universe in which galaxies did not, in fact, fill shells of increasing radii uniformly, but rather one in which galaxies appeared to form a vast, weblike structure, with great voids similar to soap bubbles marking boundaries of the voids.

This result from Mount Hopkins represented one of those truly revolutionary moments in which astronomers are forced to confront the reality of a Universe whose form seemed to reflect a different, far more complex order than had been imagined from two-dimensional images.

Not unexpectedly, this result triggered the advent of multiple surveys aimed at extending the initial work. Robert Kirshner, also working at CfA, tried a different approach: deep, "pencil-beam" surveys designed to probe the distribution of galaxies over a narrow angle but to distances much larger than the initial CfA survey. Kirshner's work demonstrated that along given lines of sight, the distribution of galaxy distances or redshifts showed maxima (at the presumed edges of the weblike structures found by Geller, Huchra, Davis, and Gursky) and deep minima (in the vast holes or voids between the galaxies). This finding confirmed the initial results and extended them to much greater distances. Other surveys have confirmed the work of the CfA astronomers and have extended maps of voids and webs over much greater areas and distances.

The quest now is to understand how this weblike structure of galaxies and voids originated. The clues to their origin seem to be encoded in the maps of the cosmic microwave background—the radiation that maps the structure of the Universe in the moments just after the Big Bang. These maps seem to suggest that the precursor structures to the observed webs and voids in the "nearby" Universe might already be apparent in temperature and density fluctuations present in the microwave background maps. Linking the earliest observable structures in the microwave background to the present distribution of galaxies represents one of the great challenges of modern astrophysics. Meeting that challenge may provide the clues that are critical to understanding the basic properties of the Universe in which we live, as well as its ultimate fate.

The Very Large Array in 2007.

National Radio Astronomy Observatory Very Large Array

Socorro, New Mexico

Radio-Wave-Gathering Gargantuans

The massive white radio antenna dishes of the Very Large Array (VLA) are sprinkled across the rugged landscape of central New Mexico like a science-fiction painting of an alien outpost. It is easy to see why these serene giants have been featured in theatrical productions ranging from the prototype modern science/action movie (*Contact*, starring Jodie Foster) to an album cover photo for the rock band Bon Jovi.

The Very Large Array is the crown jewel of the National Radio Astronomy Observatory (NRAO), founded in 1957 as the first "national observatory" open to all astronomers thanks to funding from the National Science Foundation (NSF); one year later, a counterpart for "optical" and later infrared astronomy was founded in Tucson to operate at Kitt Peak.

Managed locally in Socorro, New Mexico, about fifty miles to the east of the telescopes, NRAO also operates a major observatory in Green

Web site
www.nrao.edu

Phone
(575) 835-7000

E-mail
vla-tours@nrao.edu

Address
50 miles west of Socorro, New Mexico, on U.S. Highway 60

Bank, West Virginia, and a widely dispersed network of ten radio antennas similar to those of the VLA, called the Very Long Baseline Array (VLBA), which stretches across the Northern Hemisphere.

The primary value of New Mexico's VLA to astronomers is its ability to image faint cosmic sources at a variety of angular resolutions, at a sharpness determined by the separation of its multiple antenna components. Its twenty-seven parabolic dishes can be configured to work together as a single antenna by moving the individual dishes into four unique geometrical arrangements along a Y-shaped railroad track.

Each steel-and-aluminum antenna dish weighs 230 tons but can rotate as gracefully as a slow-motion ice skater. The scientific data from each of the 25-meter-wide (82-foot-wide) dishes can be combined electronically to mimic the resolution of a single imaginary antenna that would span a distance of up to 22 miles.

Combining the dishes' observations of the faint radio energy from distant planets, stars, and galaxies not only provides very sharp angular resolution, but also can yield an improvement in sensitivity (ability to detect faint sources) equal to what would be produced by a solid dish more than five times as large as a single VLA antenna.

Major discoveries made by the VLA range from the surprising detection of water ice on Mercury to the first detection of radio emission from a gamma-ray burst, the most powerful explosion ever observed in the Universe. Over the past three decades, the VLA also has made major contributions to our understanding of active regions on the Sun, the atmospheres of other stars, the mysterious central region of our Milky Way galaxy, and the physics of superfast "cosmic jets" of material that pour from the hearts of distant galaxies, fed by disks of gas and dust accreting onto a black hole (see pages 23–24).

After some successful preparatory experiments with a three-telescope system in Green Bank, NRAO astronomers and engineers sent the first formal proposal for the future Very Large Array to the NSF in 1967, which then carried the proposal forward to Congress in 1971. The site requirements for this array of giant radio telescopes were not very different from those for major optical observatories: a large flat area away from major population centers (in this case to avoid radio interference, not light pollution) and a dry climate (although most radio telescopes

The Very Large Array and the transfer railroad tracks.

operating at centimeter wavelengths can work in clouds and rain, they do better in clear weather). Other considerations included passable existing roads and a nearby small town of a size sufficient for staff and their families to live in.

The chosen site on the Plains of San Agustin in central New Mexico was first spotted by air inspection in 1966 and was the early favorite, winning favor when compared with two other finalists in southwestern New Mexico and Arizona in the final selection a year later. Work at the VLA site began in 1974, and by October 1975 the first VLA antenna was complete and used to observe a galaxy located 50 million light-years away in the constellation Virgo. The last VLA antenna became operational in 1980.

Pondering the existence of radio waves from deep space is easier to do if one remembers that these waves are simply a form of light that radiates with a much longer wavelength than the visual light seen by human eyes. The VLA studies radio waves with a wavelength from their peak to peak ranging from 7 millimeters to 4 meters. The waves are combined using a technique called radio interferometry, developed at Cambridge University in the mid-1950s.

The maximum angular resolution of the array of antennas is directly proportional to the wavelength of the radio waves, and inversely proportional to the maximum separation of the antennas. Hence, observations at short wavelengths and large antenna separations provide the highest angular resolution (smallest resolvable separation of objects). At a radio wavelength of 1 centimeter, the VLA has a resolution about equal to the Hubble Space Telescope's vision at optical wavelengths.

The radio waves are combined or contrasted in precise ways so as to boost the faint incoming signal from distant stellar or galactic objects, which is so tiny that it is almost beyond comprehension. The total energy of all the radio waves gathered in the name of astronomy is less than the energy released by a single snowflake hitting the ground.

The key technology underlying the VLA (and all multiple-aperture telescopes working together as interferometers) is the computing hardware and software algorithms that enable the signal from each of the telescopes to be synthesized into a coherent whole. Like ripples in a pond of water, radio waves can be combined constructively (from wave peak to peak) or destructively (a wave peak and a trough). The signals from each of the 27 VLA antennas are combined with each of the others to make 351 individual pairs. Since no pair exactly matches another in separation distance or orientation, each pair yields unique information about the object being observed.

The rotation of Earth during an observation of the sky adds another dimension, as each arm of the Y-shaped track rotates into the previous position of its neighboring arm in about eight hours. In effect, the VLA produces a "data cube" ultimately yielding brightness as a function of position and wavelength (or, alternatively, frequency—the speed of light divided by the distance between wave peaks, which yields the number of wave cycles per second).

The VLA can receive celestial radio waves in one of eight different frequency bands and its receiving system can switch among these frequency bands in about twenty seconds, making it a very powerful

The Very Large Array in snow in 2004.

instrument for observing a given source at many different frequencies nearly simultaneously.

NRAO engineers have worked for years to produce better and better radio receivers with lower and lower instrumental noise. An image that took more than twelve hours to construct in 1980 now can be assembled in a matter of minutes, thanks to improved electronics and data-processing routines. The VLA is a truly international facility, having been used by 2,500 astronomers from around the world (and counting), who have carried out more than 10,000 scientific investigations.

The VLA remains the most flexible and widely used radio telescope in the world, with upgrades to its receivers, computers, and data transmission under way that will create the "Expanded Very Large Array" (EVLA). The EVLA will increase the number of distinct spectral "channels" (narrow frequency or wavelength "bands") that can be observed from 512 to more than 4 million, thereby enabling much more detailed and sensitive analyses of the motions within cosmic sources (such as distant galaxies).

The VLA soon will have a powerful cousin in the high desert of northern Chile—the planned Atacama Large Millimeter Array (ALMA), a joint project between the United States, Canada, Europe, and Japan.

In the meantime, the VLA will continue its work for many decades. Just don't expect, as depicted in *Contact,* that an astronomer will discover signals from intelligent alien life using some portable headphones plugged into one lonely VLA antenna!

For the Public

The VLA is located 50 miles west of Socorro on U.S. Highway 60 (Socorro is about 75 miles south of Albuquerque, New Mexico). From U.S. 60, turn south on NM 52, then west on the VLA access road (166), which is well marked. Signs will point you to the visitor center.

The visitor center is open every day, except major holidays, from 8:30 A.M. to dusk. As you enter, a sign will point you toward the theater, a good place to begin your tour. The nine-minute video introduction to the VLA provides a digestible overview of radio astronomy, the technique of interferometry, and the facility. The exhibits are not elaborate or flashy, but show great care in presenting accurate information in an accessible, readily understood way. A silent five-minute video demonstrates how the antennas are moved. A small radio telescope dish located just outside the back window is usually set to actively track the Sun and to produce an image of the Sun's surface.

Near the back door, you will find a brochure to guide you along the walking tour that takes you past the outdoor Whisper Gallery and other educational exhibits to the base of one of the 230-ton antennas. From there, you can walk to the observation deck for a view of the array itself. The walking tour returns you to the visitor center and its relatively new gift shop, which opens at 9:00 A.M. and closes at 4:00 P.M. during the winter and 6:00 P.M. during the summer.

Major open houses featuring telescope tours and special presentations are offered the first Saturday in April and first Saturday in October, usually on the same days as open houses at the nearby Trinity site, where the first atomic bomb was detonated in July 1945. Educational groups can request tours by e-mailing vla-tours@nrao.edu.

The Very Large Array at twilight, facing southwest with the visitor center in the background.

For Teachers and Students

NRAO is a participant in the summer Research Experiences for Undergraduates (REU) and Research Experiences for Teachers (RET) programs funded by the NSF. It also conducts a two-week summer workshop called Radio Astronomy for Teachers, which is a two-credit course primarily for high school teachers presented through the Master of Science Teaching program at New Mexico Tech University in Socorro. The dozen teachers who attend receive lectures, participate in hands-on activities, and use the small radio telescope at the VLA visitor center and the two-element instructional interferometer at the Etscorn Campus Observatory on the campus of New Mexico Tech, operated by the New Mexico Tech Astronomy Club, to test their skill at observing.

New Mexico Tech also partners with the University of Cambridge and several other universities in the Magdalena Ridge Observatory (atop the Magdalena Mountains west of Socorro), which includes a fast-slewing 2.4-meter (95-inch) telescope dedicated in October 2006 and a planned optical-infrared interferometer. The site (at 10,600 feet in altitude) is open for public tours and amateur telescope observing (along with the VLA) during the annual Enchanted Skies Star Party in late September/early October.

A Talk with Rick Perley

National Radio Astronomy Observatory Very Large Array

Rick Perley grew up in British Columbia, the son of a banker who was transferred every few years to towns and cities all over the province. Upon reaching his third year in honors physics at the University of British Columbia (UBC) in 1967, he found a summer job working in the radio astronomy research group of the Canadian National Research Council in Ottawa.

This job featured a trip to the Algonquin 150-foot (46-meter) radio telescope, located in a wilderness park in Ontario, which started both his astronomy career and his passion for canoeing. Back at UBC, and forced to select a research focus, he picked radio astronomy thanks to his summertime experience and a new UBC professor, Bill Shuter, who was looking for students.

After two more years at UBC, which often found him distracted from academic pursuits by skiing and canoeing, he realized that he had to relocate to become serious about being a scientist. The University of Maryland had a highly regarded radio astronomy program, whose staff included Bill Erickson, who became his adviser and mentor. This led to a lot of time digging ditches and repairing cables in the southern California deserts to install and operate the Clark Lake Teepee-Tee Telescope, a low-frequency radio telescope array that Erickson was building for solar observations and other low-frequency studies.

Perley's doctoral thesis examined very-low-frequency radio emissions from galaxy clusters, including some of the lowest frequencies ever applied to the technique of synthesizing a signal from multiple radio dishes. Not long after completing it, he was selected as the first postdoctoral student at the Very Large Array.

Rick Perley, National Radio Astronomy Observatory.

What are your first memories of the Very Large Array?

I was the first postdoctoral student to join the staff of the VLA. I arrived with my wife, Peggy, in October 1977. We had naively imagined that the streets of Socorro were not paved and would be dirt. We were quite delighted to find out otherwise! Shortly after arriving, we drove out to the VLA site to get our first view, and even though it was a cloudy and dreary day, it was quite a sight.

The array had only nine of its eventual twenty-eight total antennas (including the spare). Most of the antennas did not work well yet. But the VLA had just passed the Cambridge, England, 5-kilometer synthesis array in terms of the maximum spatial resolution that it could provide, so it was a fantastic and lucky time to be a postdoc in radio astronomy. The capabilities of the VLA for resolving radio sources were unmatched and everything it did was arguably new. It was an exciting time.

What was it like during your early years?

It was a lively project in those early days. The construction crews, design engineers, were young, almost nobody over forty years old, and there were no other distractions. The whole group was focused on one thing: getting the VLA to work properly. All of us traveled on two buses, to and from work, a great way to guarantee a captive audience to discuss the problems of the day. Everybody was "rowing" in one direction.

My job was trying to develop a way of online editing of the data from the telescopes, essentially thinking of a sophisticated way to "flag" poor data to separate it from good or very good observations. But it very quickly became apparent that data came in two flavors: It was either bad or good, there was virtually nothing in between, and you could easily tell the difference.

This showed that the VLA was an excellent design, the product of extensive planning by clever engineers. So far as I can recall, this was the only thing that management ever asked me to do—I think my job was really to just use the array to do good, creative science and "make NRAO proud!"

How did your career evolve?

After three years, in 1980, the construction of the VLA had just been finished, and it was clear to me that I wanted to stay on. I became a system scientist, a job track for people with a technical bent who don't focus all their time on research papers. It was a good fit.

Even as early as the 1980s, we were thinking about what we could do to make the VLA better. Technology moves very quickly. In the late 1990s, I was asked by the director to bring a team together, go back to the drawing board, and figure out what was needed to make a big leap in capability. It was clear immediately that the improvements could not be piecemeal: There had to be an integrated plan that swept out the old and put in an entirely new system, other than the antennas and the array geometry. Antennas are expensive and there was no compelling reason to replace them.

We wanted to take advantage of new digital technologies, fiber optics, and radio receiver technology that allowed us to collect much more information than the original designers could ever have envisioned. I gave up my research career and became the project scientist for the Expanded VLA (EVLA). This was not something that could be done part time.

In 1999, we came back with a new plan, which resulted in improving the sensitivity of the VLA and extending the wavelength range over which it could observe cosmic sources. It was a "no-brainer" scientifically. For only $50 million, we multiplied the capabilities of the VLA by a factor of 10 to 100 times, which opens a vast range of new science.

We are now trying to figure out a consensus on what the radio astronomy community wants to do in the decade of 2010–2020. One idea is a high-frequency version of a new concept called the Square Kilometer Array that would take advantage of the resources and investment in the VLA.

To make a major step toward exploring new domains, we need to design an instrument to measure radio waves emanating from extraordinarily faint sources located at distances far beyond the grasp of the VLA—sources that will tell us something about the behavior of galaxies early in the history of the Universe. To do so will require ten times the collecting area of the VLA and a suite of antennas covering a distance of about 400 kilometers.

To make this project a reality, we need a technical breakthrough in how to build antennas more cost effectively; otherwise you're talking about a billion dollars. It may happen closer to 2020, but the course has to be set now. My vision is that the United States would lead it.

What are some of your favorite scientific highlights that have emerged from the VLA?

The jet in Cygnus A is the source that made my career. This unusual galaxy was predicted from its optical appearance and from other studies to have a jet, but nobody expected it to be detectable by the VLA. Some really super precision was needed to probe Cygnus A. We developed new calibration routines and used some data-processing algorithms developed by others. After numerous attempts, the jet showed up! It was a beautiful result and a great advertisement for the power of the VLA.

It is still the classic example of a really luminous radio galaxy, and truly paved the way for the kind of sensitive, high-resolution observations that continue to be the hallmark of the VLA's contribution to astronomy.

Success in this business requires not only access to new instruments, but also selecting the right objects to use a new instrument on. It's due to luck, hard work, and a little bit of good decision making along the way.

What do you hope that people will take away from a visit to the VLA?

The instrument is very grand to visit. I hope that people realize this is American science and engineering at its best. A band of young turks, many of whom preceded me, had a new technique called aperture synthesis [reconstructing high-resolution images from multiple, often widely separated telescopes used as an interferometer] and a generous funding agency in the National Science Foundation, and somehow they managed to put together a proposal for a facility that ultimately improved the state of the art by more than ten times.

By whatever magic, and a visionary leader named Dave Heeschen, this was a focused and determined group of about a dozen people that was adequately funded and politically protected. The result is a magnificent instrument on the Plains of San Agustin that still has much to contribute.

Science Highlight

The Birth of Stars and Planet-Forming Disks

Astronomers have reasonable ideas, but not yet a predictive theory, regarding the basic processes that govern how solar systems form around stars.

The central idea is that stars are born in a rotating cloud of gas and dust. Gas, along with solid grains of material comprised of silicates and other minerals, can coalesce into a protostellar core, which subsequently begins to collapse when the force of gravity of the material within its center region exceeds the outward pressure of all the constant motion of material within it. This core consists roughly of 100 parts gas (mostly hydrogen, with traces of other elements) and 1 part dust grains.

The collapsing, rotating core forms a central stellar "seed," surrounded by a comparatively thin disk. This seed/disk system is continually fed from the material in the rotating core. The orbiting disk plays a crucial role in two ways: First, in transporting material from the core to the central seed, allowing the seed to eventually (over several hundred thousand years) reach a mass comparable to that of the Sun—a blink of an eye in cosmological terms. Secondarily, the orbiting disk provides the material from which planets can form.

Following the depletion of material from the rotating natal core, the now isolated, almost mature star/disk system begins to evolve. Micron-sized dust grains (about the size of a grain of beach sand) that are embedded within the largely gaseous disk begin to settle toward the middle of the disk's plane, much as dust raised by a windstorm eventually settles earthward, clearing the air. This dust settling toward the midplane results in an increasing density of solid material, which in turn leads to frequent collisions among the micron-sized grains. Simulations both in the computer and in the laboratory suggest that once these smaller grains settle to the disk midplane, they quickly form larger grains and, soon thereafter, kilometer-sized bodies (within several hundred thousand years).

Collisions among these football-stadium-sized (and larger!) bodies, called planetesimals, can in turn produce even larger entities: solid planetary cores the size of the Moon, Earth, or even ten times larger. In turn, if there is a sufficient amount of gas and dust remaining in the disk in the frigid outer regions, gas giant planets comparable in mass to Jupiter can be assembled as massive solid cores begin to accrete gas from the surrounding disk—resulting over time in the accumulation of a huge envelope of gas.

Both the range of planetary masses and the orbital distances at which planets of a particular mass can form depend on the total amount of gas and dust in the disk and how it is distributed within the disk. For example, very-low-mass disks are unlikely to form Jupiter-sized objects because there is not enough material available. Higher-mass disks might form them easily and quite early in their evolutionary development.

Moreover, the mass of material remaining in the disk following the formation of the major planetary bodies can also significantly affect the ultimate distribution of the orbits of the young planets via their ongoing mutual gravitational interaction. In some systems, these disk-planet gravitational interactions lead to rapid migration, particularly by the giant planets (mostly inward, though sometimes outward—this is one reason that most of the planets astronomers have found so far are massive ones that orbit near their parent stars). In turn, such migration can significantly alter the orbits of less massive planets and, in some cases, expel them from the forming solar system! Such might have been the fate of Earth and its sister planets Mars and Venus if a Jupiter-mass planet migrated inward during the early formation phase of our solar system.

Determining fundamental initial properties of the disk is an essential first step in understanding what kind of planetary system is likely to form: how massive the most massive planets might be, whether they migrate (and if so, in what direction), and what the final distribution of orbital properties of surviving planets might be. In turn, statistical studies of large numbers of disks can help determine whether solar systems such as our own are likely common or rare.

The ability of the Very Large Array (VLA) to produce high-resolution radio "images" provides a unique tool for probing these processes, even for disks located hundreds of light-years from Earth. The VLA's power to carry out such observations was entirely unanticipated when it was constructed nearly thirty years ago. However, specially developed receivers built via a collaborative program with Mexican astronomer Luis-Felipe Rodriguez allow the VLA to sense radio wavelengths as short as 7 millimeters (the width of an infant's finger).

Over the past decade, this capability has provided a tool to map planet-forming disks, measure disk masses, and track the beginning stages of the planet-formation process. Observations by Luis Rodriguez, Paul Ho, and David Wilner, among others, have provided resolved images of disks around a number of forming stars and measured the distribution of dust embedded within the disks by mapping the brightnesses of the disks, as the different-sized dust grains are heated to different temperatures by their central stars. By observing disk brightness at a variety of wavelengths between 7 millimeters and about five times longer (3.5 cm), they have been able to deduce the total masses of disks and to determine that grains have indeed begun to grow—the first step in forming kilometer-sized planetesimals and, later, planets.

Similar studies using interferometer arrays specifically tailored to making observations at wavelengths as short as 1 millimeter have provided clear measurements of disk rotation speeds as a function of distance from the disk center. These data are beginning to provide astronomers with the "demographics" of planet-forming disks, and thus the tools to understand what kinds of planets are likely to form, with what frequency, and what kinds of disk architectures are most common.

VLA observations have also provided a fascinating glimpse into the effects of environment on the evolution of planet-forming disks. In particular, VLA observations carried out in the late 1980s by Ed Churchwell and collaborators led to the discovery of plumes of ionized hot gas called plasma in the regions surrounding young stars born in the Orion Nebula within the past million years. Churchwell speculated that these "plumes" were manifestations of outflows emanating from circumstellar disks heated from the outside by the strong ultraviolet radiation belched from the massive stars located at the center of the Orion Nebula. The "skins" of these externally heated disks become so hot that the atoms at the disk surface are energized enough to move at speeds in excess of the escape velocity from the star-disk system, like a mini rocket ship leaving Earth orbit.

Dramatic confirmation of Churchwell's theory was provided by images taken by Robert O'Dell (Vanderbilt University) and collaborators using the Hubble Space Telescope. When combined with spectroscopic observations carried out with Hubble, these data showed that material was being "photoevaporated" from planet-forming disks at rates sufficient to reduce the disk mass in some cases by a factor of 10 in a matter of a million years.

The harsh environment in regions like the Orion Nebula can thus remove material from planet-forming disks quickly enough to "neuter" the production of extrasolar Jupiters. For those disks that do manage to form Jupiter-sized bodies, the rapid loss of mass in the outer parts of the disk may preclude significant inward migration of the young gas planets.

The significance of these observations lies in the fact that the majority of stars in the Milky Way are born in regions like the Orion Nebula. Does the fact that Jupiter and Saturn did not migrate inward imply that our solar system formed in a region like Orion? Could this famous target of telescopes large and small be a flashback to our ancient past?

The measurements of planet-forming disks now barely possible with the VLA provide a taste of what will be possible when its millimeter-wave analog, the Atacama Large Millimeter Array (ALMA), starts work in Chile in the coming decade. A collaboration among the United States (led by NRAO), Canada, Europe, and Japan, ALMA consists of up to eighty 12-meter-wide (39-foot-wide) radio dishes spread out across the high desert of the Atacama Region of northern Chile, one of the driest and most remote areas on Earth. ALMA builds on the experience gained with the VLA in combining images from multiple, widely spaced "small" antennas to synthesize a high-resolution image, this time at wavelengths twenty times smaller than those accessible to the VLA.

ALMA will enable observations at wavelengths between 0.35 and 7 millimeters, thus extending the range of the VLA into a region that will enable observations of the early Universe, 12 billion years ago. At this ancient time, much of the energy from stellar objects has been stretched (or "redshifted") to longer wavelengths because of the expansion of the Universe. ALMA will also enable more detailed studies of the physics, chemistry, and kinematics of protostellar clouds and disks—the sites of star and planet formation. At its maximum resolution, ALMA will be capable of imaging a region as small as 1 astronomical unit (the distance from Earth to the Sun) at the distance of the nearest star-forming regions, 400 light-years away.

With its ability to observe emission from both glowing dust grains and excited atoms and molecules, ALMA will be able to trace the shapes, velocities, and chemistries of disk gas and dust. In disk systems that are favorably aligned with our view from Earth, ALMA should be able to see indirect evidence of forming planets from "gaps" produced as a result of gravitational interactions between protoplanets and surrounding disk material. These observations will be able to tell us directly what kinds of planets form in what kinds of disks, where they form, and when (based on optical studies of the luminosities and temperatures of parent suns).

In a decade's time, it is likely that we will have developed both a rudimentary predictive theory of planet formation and a much deeper understanding of whether systems like our solar system are common or rare, answering questions raised for centuries by scholars and philosophers alike.

An aerial view of Sacramento Peak Observatory with the Evans Solar Facility in the foreground, the Dunn Solar Telescope, and the Hilltop Dome Facility in the background.

The Observatories of Sacramento Peak

Sunspot, New Mexico

Solar Pioneers and the Greatest Sky Survey to Date

The Apache Point Observatory and the National Solar Observatory (NSO) at Sacramento Peak (Sac Peak) offer a special two-for-one chance to experience the vast progress of astronomy in the latter half of the twentieth century.

The observatories stand in the Lincoln National Forest in central New Mexico's Sacramento Mountains, a thirty-five-minute drive up one main road from Cloudcroft, a charmingly compact small town sixteen miles east of the iconic city of Alamogordo. The peak is surrounded by a variety of scenic hiking trails and expansive views of the Tularosa Basin, including world-famous White Sands National Monument. The village at the peak is known as Sunspot, a name chosen by James Sadler, the U.S. Air Force officer in charge of constructing the observatory, who went on to international scientific fame as an extreme-weather meteorologist at the University of Hawaii.

Web site
www.nso.edu/visitors.html

Phone
(575) 434-7190

Driving Directions:
nsosp.nso.edu/pr/road-directions.html

The observatories of Sac Peak present an appealing mixture of uniquely shaped solar telescopes that work during the day, the nation's most active remotely operated midsized optical telescope for nighttime astronomical research, and the single greatest wide-area survey telescope in the world.

The roots of this historic site lie in the once-mysterious, still-unpredictable power of the Sun to disrupt radio communications and related electromagnetic waves in Earth's upper atmosphere. The U.S. military recognized the serious implications of this interference during World War II, so attempting to understand and forecast these phenomena became a natural focus of work for researchers supported by the air force.

In particular, the dry pristine skies above Sunspot enable telescopes to look at the tenuous hot plasma extending well above the Sun's surface (the solar corona) and its surface activity (cold sunspots and hot active regions) in great detail. More than five decades after its founding, the National Solar Observatory at Sacramento Peak remains a world leader in high-resolution studies of our nearest star.

The hardy men who founded the solar observatory hailed largely from Colorado, but the rugged isolation of its location caused these men to include their families as residents from its earliest outpost-flavored days.

The first key figure was Leadville, Colorado, native Rudy Cook, who met solar scientist Walter Orr Roberts on a high school class trip during World War II. Cook served a stint in the army, then contacted Roberts during his postwar search for employment. Roberts was in the process of founding the National Center for Atmospheric Research in Boulder. Using funds from the air force, Roberts and solar astronomer Jack Evans hired Cook to scout out and establish a new high-altitude solar observatory station in the Sacramento Mountains in the summer of 1947.

After a few initial reconnaissance trips to the isolated 9,200-foot peak using a military weapons carrier as transport and a railroad boxcar as living quarters, a crude road was constructed. Cook went back to Colorado and returned with his wife and three-year-old daughter in 1948. A second pair of newlyweds, Lee and Rosemary Davis, were also betrothed that fall and also began their new life together on Sac Peak.

Local cattle ranchers Jean and Bill Davis (unrelated to Lee) had a self-sustaining settlement about eight miles from the peak that saved the early crew of the observatory more than once, according to the fascinating anecdotes of Joanne Ramsey, wife of the third major staff member, Harry Ramsey. Early life was a rugged, often invigorating, sometimes muddy mess, exemplified by premanufactured Quonset huts and gasoline-powered wringer washing machines. Living at the edge of oxygen deprivation in such rough conditions sometimes led to some petty disputes among the residents, but the day-to-day challenges of life on the peak let these incidents pass quickly and led to incredible personal bonding among the early residents.

By 1952, a series of redwood houses and a combination cafeteria/administration building/community center had been built for staff and their families; the buildings survive and continue to serve staff well today. These first signs of any "luxury" were followed by a post office stop in Sunspot established in 1953, a milestone that signified the arrival of mainstream civilization.

The first permanent solar telescope was built in a grain bin ordered from a Sears catalog (the same handy sourcebook used to purchase the first garage workshop on the rugged mountain). The grain bin was rebuilt by air force machinists to rotate and to have a door. The plainly named Grain Bin Dome began work in 1950 with a 6-inch (0.15-meter) telescope designed to look at large active tendrils of hot gas on the limb (edge) of the Sun known as solar prominences. From March 1951 through 1963 daily images seeking new solar flares were taken from the Grain Bin, until its duties were finally handed off to the newly built Hilltop Dome Telescope.

The Grain Bin continued to be used for coronal observations until 1969. It survived a period of disuse (including a roof collapse from heavy snow) to be revived today as the home of a small telescope for nighttime viewing by mountain residents.

Another Sac Peak facility that breaks the mold of the typical research telescope shape is the Big Dome—also known as the Evans Solar Facility, in honor of solar astronomer Jack Evans. Completed in 1953, this squat pyramid-like building resembles a visual twist on a 1950s flying saucer.

The telescope on the large spar (or column) inside the Big Dome is mostly used as a coronagraph. It helps scientists study the faint outer layers that lie beyond the easily visible solar disk (the corona) of the Sun by blocking the light from the bright circular body of the Sun using a round disk inside the telescope—thereby permitting observations of the shapes and physical characteristics of the hot plasma comprising the solar corona. In effect, a coronagraph creates its own solar eclipse whenever one is needed. It is only effective as a science instrument in clear, thin air free of large amounts of dust, which is why the Evans facility is located atop Sac Peak.

The Big Dome Telescope does not sit straight on the big concrete spar. Instead, it is mounted at an angle of 35 degrees relative to the horizon. If you were to draw a line from the spar through the telescope and extend it into the sky, then you'd end up near the North Star, around which the whole sky (including the Sun) seems to rotate once a day. This means that the telescope can simply follow the Sun along its daytime track in the sky by rotating in one direction only, parallel to the imaginary line projected outward to the North Star, using a motor that works at constant speed. The machinery of the Big Dome facility shows that this elegantly practical configuration, called an equatorial mounting, works as well during the daytime as at night.

A small building attached to the Big Dome uses a movable two-mirror system called a coelostat to feed a stationary image of the Sun to a suite of instruments that monitor the Sun all day for transitory activity such as energetic solar flares that eject hot plasma outward from the solar surface. A small observation room at the left-hand side of the Big Dome (as seen from the road) is open to the public.

The other multifunction facility in Sunspot is the Hilltop Dome. From the outside, it resembles a mini version of the 200-inch dome at Palomar, with a service building attached to the side. Inside, the dome houses another large, equatorially mounted spar to track the Sun, this one with eight sides capable of holding a scientific instrument. The instruments are a mixture of experimental devices and longstanding workhorses that take regular images of the whole Sun in white light (basically the range of wavelengths visible to the human eye) and in

The telescope spar in the Evans Solar Facility at Sacramento Peak.

selected wavelengths that highlight the formation of solar flares. The resulting photographic record runs decades long. Recent images are displayed on one of the TV monitors for the public inside the lobby of the Dunn Solar Telescope.

The Dunn Telescope is the signature building on the solar side of Sac Peak. Visible from White Sands and other distant points, the Dunn Telescope is a striking blend of the angular beauty of an ancient pyramid with white-painted modern scientific functionality. The Dunn is 136 feet tall and extends nearly 230 feet more than that underground. The main function of all this external structure is to support an innovative entrance window at the top, where two mirrors guide an image of the Sun down and through a 329-foot-long tube that has been evacuated of air: essentially a vacuum chamber.

The vacuum inside the telescope tower serves to pass the image of the Sun through a turbulence-free tunnel, feeding the pristine image to a variety of imaging cameras and spectrographs to dissect the incoming sunlight into its component colors. In recent years, the staff of the National Solar Observatory and collaborators at places like the New

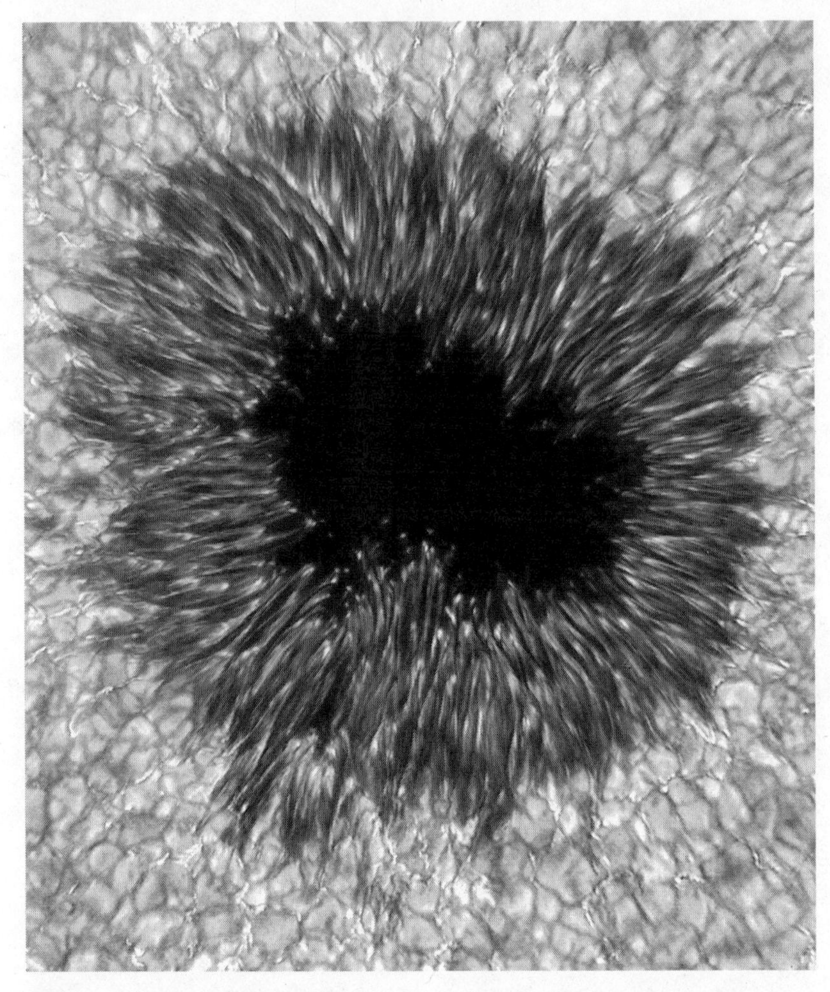

High-resolution image of a sunspot produced with a modern CCD camera attached to the Dunn Solar Telescope's adaptive-optics system. The dark black area (the umbra) is roughly two Earths in size.

Jersey Institute of Technology have developed some of the most advanced adaptive-optics systems in the world for solar science—enabling imaging and spectroscopy of the Sun at resolutions comparable to that achievable in space by correcting for the distorting effects of turbulence in the upper atmosphere in real time.

Originally known—for obvious reasons—simply as the Vacuum Tower Telescope, it was renamed in 1998 for Richard B. Dunn, a legendary figure in solar astronomy who conducted observations for his doctoral thesis at Sac Peak in 1951 and caught the eye of Jack Evans, who was the director of the observatory at that time. Dunn's specialty was making observations of solar prominences and other surface features at the sharpest-possible resolution. Dunn used the telescope to make historic observations of transient jets of solar gas known as spicules and thin filigrees of magnetic fields in the Sun's photosphere at the edge of the telescope's capability to resolve separated features, also known as the diffraction limit. (The diffraction limit is proportional to the wavelength of light being observed divided by the diameter of the telescope. The bigger the telescope, the smaller the diffraction limit and the smaller the distance between objects that are at the limit of the telescope's resolution.)

The primary mirror of the Dunn Telescope is 1.6 meters (64 inches) in diameter, comparable to the bigger nighttime telescopes in the world when it was dedicated in 1969. The telescope is equipped with instruments that work in visible light and into the near-infrared region of the spectrum (wavelengths between 1 micron and 2 microns). It can produce an image of the Sun that is 20 inches in diameter with enough sharpness to resolve features on the surface of the Sun as small as 90 miles in diameter when magnified. This is a relatively small area, considering that the Sun is a ball of incredibly hot gas and plasma that is 9,600 times wider than that, with a volume that could hold more than a million Earths.

The central vacuum tube and everything connected with it—a mass of 200 tons—must rotate to compensate for the apparent motion of the Sun (as seen from the telescope) as our nearest star completes its arc across the sky each day. Instead of being supported from the bottom, as one might imagine, the entire structure is hung from a bearing near the top, which resides in a large tank of mercury. The balance of the system is so fine that a person can rotate the tube pushing with one foot on the main floor and one on the turntable. The Dunn Telescope remains at the forefront of high-resolution imaging of the Sun a half-century after it was conceived.

The National Science Foundation took over full responsibility for operating Sac Peak from the U.S. Air Force in 1976, formally changing its primary character to pure research, though a core air force staff remains today. The Dunn and Evans telescopes continue to enable studies of sunspots, active regions, flares, and the solar corona, toward the overall goals of understanding the basic physical processes that control energy transport throughout the Sun and monitoring (and beginning to forecast) the processes that affect the transport of energetic particles from the Sun to Earth.

In the early 1980s, Sacramento Peak astronomers such as Jacques Beckers recognized that the time was ripe to add nighttime telescopes to the mountain. Site testing with some of the solar telescopes verified the high quality of the conditions at night, thanks to the naturally smooth flow of air rising upward from the Tularosa Basin below. New Mexico State University and the University of Washington in Seattle hatched plans to install a 2-meter (79-inch) telescope, which later drew interest from the University of Chicago, Princeton University, and Washington State University. This group, later augmented by Johns Hopkins University and the University of Colorado at Boulder, formed a partnership called the Astrophysical Research Consortium (ARC).

Frustrated by insufficient observing time on national facilities, ARC engaged Roger Angel of the University of Arizona to use his new mirror-making laboratory to construct a 3.5-meter (138-inch) mirror by a creative new technique called spin casting. Developed by Angel and his team, this technique used a rotating oven to melt glass over a form of ceramic inserts. The spinning motion gives the mirror its basic parabolic shape; once it has cooled and hardened, the ceramic cores are blasted out with high-pressure water, leaving just the mirror blank, ready for polishing. The resulting hollow honeycomb spaces allow the mirror to cool much more quickly to match the surrounding air temperature, greatly improving the quality of images that it can produce.

The ARC 3.5-meter telescope has been a pioneer in the field of remote observing, where the astronomers remain at their home institutions and control the telescope and its instruments from a distance. This mode, with its obvious appeal in terms of saving travel time and costs, had been explored at places like Kitt Peak years earlier, but ARC was the first to make it work effectively, thanks in part to major advances in control technology and the power and speed of computer networks.

The 3.5-meter telescope has made a variety of discoveries, many recent ones in combination with its more famous neighbor on the mountain, the Sloan Digital Sky Survey (see below). For example, in 1999, a near-infrared camera mounted on the 3.5-meter telescope carried out observations of multiple deep-space objects first imaged by the Sloan survey, which led to the discovery of some of the most distant galaxies then known.

The 3.5-meter has also been used to carry out observations of cool, gaseous, starlike objects known as brown dwarfs, which are cooler than stars and hotter than planets. Brown dwarfs are defined as having a mass too low (less than 8 percent of the mass of the Sun) to support nuclear reactions at their centers. Astronomers using the ARC 3.5-meter found the first brown dwarf to exhibit evidence of the methane molecule, which is characteristic of extremely cool objects (with only one-tenth the surface temperature of the Sun) whose masses are only a few times those of the most massive planets.

In 2007, the 3.5-meter began science observations with the Apache Point Observatory Lunar Laser Ranger Operation (APOLLO), which makes use of the retro-reflectors installed on the Moon by the NASA Apollo mission astronauts in the late 1960s and early 1970s. APOLLO sends laser pulses toward the lunar retro-reflectors and makes precise measurements of the time it takes for the return signal to arrive back to Apache Point in order to measure the relative distance of the Earth-Moon system to a precision of 0.04 inch over a distance of 240,000 miles.

This remarkable capability enables astronomers and physicists to track the slow, nearly imperceptible outward motion of the Moon's orbit away from Earth and to make sensitive tests of the fundamental properties of gravity, such as whether Newton's gravitational constant is indeed a constant or might change with time on a scale of one part in one thousand billion.

However, the most widely known and referenced research to emerge from Apache Point has flowed primarily from its other major telescope and a tremendously vast survey it conducted, the Sloan Digital Sky Survey.

The Sloan Digital Sky Survey 2.5-meter telescope with the Sacramento Mountains in the background.

The 2.5-meter (100-inch) telescope used for the Sloan Sky Survey is not particularly advanced in itself. However, it is equipped with a 125-megapixel digital camera that enables deep imaging of the sky over a huge area (27 lunar diameters) as well as a spectrograph that allows simultaneous observations of hundreds of targets, which can be selected for further detailed study following analysis of deep images. Somewhat unusual among large research telescopes, it is housed in a roll-off building—suspended over the edge of its mountain cliff 40 feet above the ground—that slides back at night to expose the facility to the open night air.

But the true uniqueness of the Sloan rests on its landmark effort to observe a huge swath of the northern sky with great fidelity and consistency. Using creative teamwork and funding support from the Alfred P. Sloan Foundation (along with taxpayer contributions from the

The Sloan telescope enclosure building rolled over the telescope, with its instrument room (below), juts out from the mountainside. The facility was designed this way to provide the calmest and coldest possible air above and around the survey telescope.

National Science Foundation), a coordinated group of three hundred astronomers and engineers at twenty-five different institutions has used the Sloan Telescope to extend the legendary Palomar Sky Survey to a new frontier.

The first five-year round of the Sloan survey, 2001–2005, took images of 300 million celestial objects and spectroscopic measurements of 800,000 galaxies, 300,000 stars, and 104,000 quasars (the bright cores of active galaxies that contain black holes). These surveys have deepened our understanding of how galaxies are arranged in space (see pages 89–91) and how they evolved from approximately a billion years following the Big Bang to the present.

The Sloan Telescope has also played a major role in probing the population of stars in our own galaxy, searching for rare brown dwarfs, and charting the distribution of various "stellar populations" (stars spanning a range in chemical composition, age, and kinematic properties) in the Milky Way—a study that is key to understanding how our parent galaxy was assembled.

Astronomers proposed, and received funding for, a second round of experiments with the Sloan Telescope, which ran from 2005 to 2008. Sloan-II was aimed at refining the initial survey, filling in some gaps, and extending the range of time over which variable objects are observed.

A third round of experiments, now under way, is slated to run until mid-2014. Sloan-III includes four coordinated surveys that are studying the expansion of the Universe, mapping the outer reaches of our Milky Way galaxy, quantifying the nature of more than 100,000 stars using infrared light observations to peer through the dust that usually obscures them, and searching 10,000 stars for evidence of planets around them. Sloan-III will cost approximately $7 million, a large sum but only a fraction of the cost of a new telescope.

The Sloan team has also undertaken some creative public outreach, with a pathfinding online citizen-science experiment called Galaxy Zoo (www.galaxyzoo.org) that has enabled more than 200,000 people to participate in research by visually classifying the types of galaxies in the Sloan Sky Survey images. When averaged over the millions of characterizations of the total sample, the Galaxy Zoo campaign produces scientifically useful statistics—innovative enough to trigger follow-up observations at places like the WIYN 3.5-meter telescope on Kitt Peak. Galaxy Zoo has been such a successful experiment in citizen science that a more elaborate second-generation version known as "Zoo 2" was released in early 2009.

For the Public

The National Solar Observatory at Sacramento Peak lies in New Mexico at the southern end of NM Scenic Byway 6563, about 18 miles south of Cloudcroft (on NM 82) and 40 miles southeast of Alamogordo (on NM 70 and 54), in the village of Sunspot.

The drive from Alamogordo in the Tularosa Basin to Cloudcroft in the Sacramento Mountains involves a curving, 16-mile climb spanning 4,500 feet in altitude, and thus takes more time to navigate than a comparable distance on a flat highway. Expect the drive to take an hour from Alamogordo to Sunspot and about thirty minutes from Cloudcroft to Sunspot. Sunspot has no public gas station, grocery store, or restaurant, so make sure you have enough gas and food to return to Cloudcroft. Water, soft drinks, and snack food items are available at the Sunspot Visitor Center.

From Cloudcroft, take NM 130 East (the junction with NM 82 lies at the western edge of town) and drive about 2 miles to the junction with Sunspot Highway (NM 6563). Follow the highway all the way to its end, about 14 miles.

In 2008, the NSO built a 1:250-million scale model of the solar system along this route. Eight signs along Sunspot Highway mark the orbits of the planets, and models at Sunspot depict the Sun (an 18-foot dome on this scale) and the planets (Earth is just 2 inches wide). As you arrive, note two tiny signs: PENUMBRA ("almost" a sunspot) and UMBRA (the center of a sunspot). Once inside Sunspot (through the stone gate), take the first turnoff to the left (marked "Visitor Center") and park in front of the center.

The Sunspot Visitor Center and Museum first opened in July 1997. The center is a collaboration between NSO/Sacramento Peak, Apache Point Observatory, and the U.S. Forest Service. It contains a variety of exhibits about astronomy and the tenants of the mountain, free maps for the self-guided walking tour, and a well-stocked gift shop. Perhaps the highlights of its collection are found outside near the entrance, where a beautiful sundial and an ancient model of the celestial sphere known as an armillary sphere are displayed.

The Sunspot Visitor Center.

The visitor center is accessible by wheelchair and is the start point and end point of guided and self-guided walking tours of the observatory. It is open every day, except major holidays, from 9 A.M. to 5 P.M. (weather permitting). The walking tour and restrooms are available for visitors at all daylight times throughout the year (weather permitting). Call ahead if necessary to (575) 434-7190.

Guided tours of the solar observatory are offered daily for a small fee at 2 P.M. Reservations are not required.

Groups of at least ten visitors, of all ages, can arrange for private guided tours any day of the week, year-round (depending on the availability of a guide). Beyond the general walking tour and visits to the Dunn Telescope (about forty-five minutes each), the standard private guided tour usually includes a slide show of comparable length.

For more information and reservations, call (575) 434-7003 or fill out a request form at nsosp.nso.edu/pr/tour-form.html.

The visitor center includes exhibits on Apache Point Observatory, which does not offer regular tours but welcomes visitors to stroll its grounds between 7 A.M. and 5 P.M. Special tours can be arranged by contacting the observatory at (575) 437-6822. For the latest information on Apache Point, see www.apo.nmsu.edu.

For Teachers and Students

The National Solar Observatory maintains an active Research Experiences for Teachers (RET) program, where high school teachers work with NSO staff scientists over the summertime to develop classroom exercises and to bring the flavor of research back to their students.

It also offers online learning activities for the middle school and high school levels, based on more than twenty-five years of measurements of the Sun's magnetic field—see the Data and Activities for Solar Learning (DASL) project at eo.nso.edu/dasl.

A Talk with Steve Keil

Director, National Solar Observatory

Emblematic of the long-term hold that Sacramento Peak Observatory holds over many of its residents, Steve Keil came to the peak as a summer student in 1973 and 1974, returned again for postdoctoral work via the University of Colorado in 1975, then became a member of its U.S. Air Force scientific staff through 1999, when he became director of the National Science Foundation's National Solar Observatory (NSO). Today, Keil has expanded his research on the interaction between the magnetic fields near the surface of the Sun and the convection of its heat to include the complexities of the outer region of the Sun's atmosphere known as the corona, as he simultaneously guides the NSO toward construction of an advanced solar telescope four times larger in diameter than any in operation today.

Steve Keil, director, National Solar Observatory.

What are the earliest roots of Sacramento Peak Observatory?

Right after World War II, the air force realized that solar storms were causing interference in radio communications, especially long-range communications that went over Earth's poles. So they became very interested in trying to understand the causes. The Air Force Cambridge Research Lab got Harvard College involved, and it hired the High Altitude Observatory in Colorado to find and establish a site. The air force had a solar coronagraph at Climax Observatory in Colorado, which was a difficult place to observe from during the winter. They were looking for a dust-free site and, at the time, the Southwest was going through a wet period, so New Mexico looked very attractive to relocate the telescope from Climax.

In the early 1950s, the air force built the world's largest coronagraph at the time and installed it at what we now call the Evans Solar Facility on Sac Peak, figuring that if you could understand the behavior of the Sun's outer atmosphere, you could understand how these storms escaped to influence interplanetary space. Then they decided they wanted a predictive capability, which led to the Hilltop Dome and the flare patrol instruments. This helped people realize that, because of the Sun's rotation, the flares that affect Earth appear first on the western hemisphere of the Sun.

The Dunn Solar Telescope is probably the most iconic facility at Sac Peak. What led to its unique design?

In the late 1950s and early 1960s spectroheliographs [instruments to image the Sun at a particular wavelength, or color] were being employed to measure the surface of the Sun and see active regions like sunspots, and then surface magnetograms. This led to the belief that the Sun's magnetic field was interacting with the plasma comprising the corona on smaller scales than we had been able to observe.

There were large solar telescopes at the time, but their images were bad because of heating of the turbulent air inside them. Solving that problem led to the design and construction of the Dunn Solar Telescope with its evacuated light path. The design by Dick Dunn became sort of the standard. It enables you to get the entrance of the telescope up high to avoid the natural air turbulence stirred up near the ground. You couldn't turn the telescope in traditional ways because of the big vacuum chamber, so turrets at the top became the way to direct the light. The subsequent German and Swedish tower telescopes were based on that concept. The big pyramid shape of the Dunn is essentially a way to hold up the top of the telescope with a very long focal length, which gives you a very flat image of the Sun.

Why is it still important to have ground-based telescopes in this age of very high-tech satellites?

Primarily, it's the flexibility to quickly change instrumentation to keep up with the scientific need. You can use multiple instruments simultaneously, and you can fix them more easily if they break. If you discover something new on the Sun, you can rapidly change the configuration that you have to observe it. In space, you're stuck with what you launch. Data rates are also a big thing—you can handle much higher ones on the ground, which is very important for the multi-wavelength, multi-camera observations that we do today.

The other element is money: It costs about ten times as much to get the same size aperture in space as on the ground. The disadvantage is that there is no access to x-rays or ultraviolet light, which we can get in space. It's really the combination of high-energy observations from space and high-resolution spectroscopy from the ground that drives solar science today.

What is the connection between the current telescopes at Sacramento Peak, such as adaptive-optics technology on the Dunn, and the 4-meter Advanced Technology Solar Telescope (ATST) that you hope to build in Hawaii?

When we achieved the diffraction limit of 1-meter telescopes using adaptive optics (the theoretical best image quality that can be produced for a given aperture size), it became clear that there were some very interesting features that were still unresolved. Resolving these features requires a larger telescope with a smaller diffraction limit.

The other problem is not being able to gather enough photons of light. In order to measure polarization of small features on the Sun [which is diagnostic of magnetic field direction], you need lots of photons in a shorter timeframe than the lifetime of solar features; again, that requires a larger telescope that collects more photons. Short exposures also mitigate the effects of changes in Earth's atmosphere [also known as "seeing," which results from the passage of turbulent elements of air above the telescope, with different effects at different altitudes]. A large photon count also produces a strong enough signal against the background noise of the observation to allow accurate measurements of solar magnetic fields.

So, you need more light, which is hard to believe given that it's the Sun we're talking about, and you need to be able to collect it in a much shorter amount of time.

How do solar telescopes overcome the heat generated by sunlight so that it does not melt your instruments?

So far, the technique has been to build telescopes with really long focal lengths, so you don't concentrate the heat in a small area. The Dunn Telescope gives you an image of the Sun about 20 inches across and, as a result, the light is spread out and does not raise the temperature of the focal surface by a significant amount. For the ATST, thermal control is a big cost driver. The telescope admits a much smaller field of view, and we have to cool the primary and secondary mirrors about two degrees below the ambient temperature so the heat flows down and does not produce turbulence above the mirror. Dust is also a big issue—scattered light is actually a bigger concern than the quality of the mirror. So we have designed systems to actively clean the mirror.

It seems that family life has always been a significant element of the observatory. What is it like to live on Sacramento Peak?

We've always had a real strong community group, and it's still going strong today. It's really the heart of our social interactions. They meet once a month, and have a potluck dinner once a month. They give out a scholarship every year to local high school students. Some people come for a year and leave. But if you stay for two years, you tend to stay forever.

Life at Sac Peak can be a mixed blessing, because it is isolated. Some people really prefer city life. It can be difficult for spouses to find work, unless they are teachers or self-employed. Some of the spouses have been drawn into work with Apache Point Observatory, such as helping to install the optical fibers in the spectrographic plates used in the Sloan survey, or serving as spotters for their laser. But it's a great place to raise kids—the phrase "it takes a village" is very applicable to life there. I've never locked my door in the thirty-five years that I've lived there. And it continues to be a joint facility with the U.S. Air Force, with each of us having about a half-dozen scientists based there, so in terms of scientific vigor, it's still a good place to work.

The Dunn Solar Telescope is popularly rumored to be the burial site of the alien bodies from the infamous Roswell incident of 1947. What can you tell us about this situation?

Well, it's true that the Dunn Telescope is only a three-hour drive from Roswell, New Mexico, and that we are very nearby Holloman Air Force Base, which has been known to conduct some highly classified "black" programs over the years. One of the key figures in the founding of Sacramento Peak Observatory, Donald Menzel of Harvard College Observatory, is often considered to be part of a secretive society of astronomers who offered Cold War–era advice to the military. Finally, why else would the air force dig a 230-foot hole in the ground at such a remote wooded place? Hmmm. Unfortunately, the aliens are usually out gallivanting around causing UFO sightings, and we have not seen them.

Science Highlight

The Complexity of Sunspots

Sunspots are apparently dark regions visible on the surface of the Sun. They were first studied in a systematic way by Galileo in the early seventeenth century. His pioneering observations in 1612 revealed the vast size of sunspots, their impressive persistence, and their intriguing eleven-year cycle. The largest sunspots span thousands of miles in diameter, wider than the diameter of Earth, and can sometimes last for several thirty-day-long complete rotations of the Sun.

The true nature of sunspots remained a mystery until twentieth-century measurements demonstrated that they are relatively cool regions, with temperatures about 1,500 degrees lower than the 9,800-degree temperature typical of the solar surface region, called the photosphere. The characteristics of sunspots are deduced from measurements of the relative strength of features in the spectrum of the Sun that arise from the gaseous form of elements such as iron, calcium, and magnesium. These elements absorb or "block" certain wavelengths of light, causing features known as absorption lines. First observed by Newton using a large prism, spectral absorption lines are now measured precisely using instruments known as spectrographs.

Why are sunspots cooler than their surroundings? The key to answering this question was the discovery that sunspots are regions characterized by strong magnetic fields. Powerful spectrographs on telescopes revealed the detailed shapes of the absorption lines. These shapes show "broadening" and "splitting" of individual spectral lines that are produced when strong magnetic fields thread through hot, gaseous plasmas.

From measurements in terrestrial laboratories and theoretical calculations of line broadening and splitting, it was possible for astronomers to translate measurements of line shapes and splitting observed within and outside sunspots into measurements of magnetic field strength. Astronomers at Sacramento Peak and at their sister office in Tucson that runs the National Solar Observatory telescopes atop Kitt Peak have pioneered techniques for producing regular images of the Sun's magnetic field—images produced from measurements of line broadening and splitting and available for use by astronomers and other scientists throughout the world.

The fact that sunspots are sites of enhanced magnetic field strength results in a significant decrease in the efficiency of energy transport from the hot interior regions of the Sun in the vicinity of sunspots as compared to the surrounding regions of the solar photosphere, where heat transport proceeds unimpeded. As a result, sunspots are cooler than nearby regions of the photosphere.

Sunspots have long been known to be the origin of bursts of energetic activity. The most prominent of these are solar flares. Such flares, which occur most frequently during the part of the eleven-year cycle when the number of sunspots reaches a maximum, are generated as the energy stored in the twisted magnetic field "loops" above the spots is released suddenly, launching hot plasma into space. Some of the charged particles from the released plasma can reach Earth and penetrate our protective magnetosphere, sometimes interrupting radio communications and disrupting electrical power grids. The charged particles from flares can also produce auroras, as the energetic electrons emanating from flaring events excite atoms in the upper atmosphere (such as oxygen) and create glowing sheets of light.

Sunspots have thus pointed the way toward understanding the complex interactions of gas motions and magnetic fields that transport energy from the interior of the Sun (where nuclear reactions unleash energy sufficient to heat the gas to temperatures of millions of degrees) to its surface. Gaining a deep understanding of energy transport both inside the Sun's hot plasma and across its complex outer atmosphere is vital to understanding the factors that influence the total energy per second reaching Earth as well as the spectacular "solar storms" created by flaring events. These factors are in turn crucial elements of the external forces that drive long-term climate changes on Earth, as well as far shorter-term terrestrial responses to solar activity, dubbed space weather.

Astronomers at Sacramento Peak Observatory have designed and constructed a variety of sophisticated instruments aimed at making detailed measurements of the physical properties of the outer layers of the Sun: temperatures, magnetic field strengths, and the velocity of the hot gases, both those in the deceptively quiescent regions of the solar photosphere and those around the sunspots that fascinated Galileo almost four hundred years ago.

Along with their Kitt Peak colleagues at the Tucson office, Sac Peak astronomers have carried out systematic studies of gas motions and magnetic field structure in the Sun's outer layers. Combined with international efforts to study the structure of the solar interior (via a technique analogous to sonograms) as well as hot plasmas above the photosphere extending into interplanetary space (using a variety of remote sensing and spacecraft-based measurements), National Solar Observatory astronomers and their colleagues have begun to develop an empirical and theoretical framework for understanding the basic workings of our parent star.

Further progress in sharpening our models of solar behavior and making them reliably predictive requires observations of features on the Sun at a far smaller scale than has been possible in the past.

Current solar astronomers at Sac Peak, such as Thomas Rimmele, have pioneered techniques to use adaptive optics during the daytime to compensate for the blurring effects of Earth's atmosphere, without the benefit of the bright stars or laser beams used by nighttime astronomers. Over a period of fifteen years, NSO astronomers have used the Dunn Solar Telescope to produce images of a clarity unprecedented for any solar telescope in the world. These observations have led to breakthroughs in understanding how sunspots work and, more generally, produced a deeper understanding of how energy is transported in the presence of a magnetic field—both the weak fields permeating the "quiet" photosphere and the strong fields found around sunspots. Solar scientists are now applying this experience to the design and development of the next great solar telescope, the planned 4-meter Advanced Technology Solar Telescope, to be built on Mauna Loa in Hawaii.

The Hobby-Eberly Telescope as seen from the air.

McDonald Observatory

Fort Davis, Texas

A Jewel of West Texas

The major telescopes of McDonald Observatory sit astride two neighboring peaks in the rugged Davis Mountains of west Texas, underneath some of the darkest skies in the continental United States. McDonald Observatory has a rich history that runs both eastward through the Texas plains to the campus of the University of Texas in Austin and northward to hilly Wisconsin and two of the greatest astronomers of the middle twentieth century. Thanks to the advocacy of its charismatic directors, McDonald Observatory has also long been a leader in programs to educate the public about astronomy.

Given its remote location, more than a seven-hour car trip west from Austin and still a fair drive from the nearest landmarks (Interstate 10 to the north and Big Bend National Park to the south), a trip to McDonald Observatory requires a clear commitment.

The rolling scenery, the diverse telescope facilities, and a first-class modern visitor center make the trip worth the effort. An overnight stay in nearby small-town Fort Davis (sixteen miles away) adds to an undeniable sense of splendid isolation and a flavor of the way the western United States must have felt before the age of superhighways.

Web site
mcdonaldobservatory.org/visitors

Phone
(877) 984-7827 toll free

Address
HC 75, Box 1337-VC
Fort Davis, TX 79734

The large telescope domes of McDonald Observatory. In the top left corner, the Hobby-Eberly Telecope dome sits atop Mount Fowlkes. In the foreground, the dome of the Otto Struve Telescope sits at left and the Harlan J. Smith Telescope at right, atop Mount Locke.

McDonald Observatory is operated by the University of Texas at Austin, which manages the observatory from offices on campus, more than four hundred miles away. The observatory owes its roots to the lucky intersection of a large donation to the university by wealthy bachelor banker William J. McDonald and the desire of the University of Chicago and two leading astronomers—Otto Struve and Gerard Kuiper—to find a dark site that would improve upon their existing facilities at Yerkes Observatory in snowy southern Wisconsin.

The University of Texas and the University of Chicago combined their resources to build an 82-inch (2.1-meter) telescope on 6,791-foot Mount Locke which became the second-largest telescope in the world when dedicated in 1939. Kuiper utilized it immediately for a landmark study of the chemical composition of four hundred stars and, later, to discover small moons around planets Uranus and Neptune.

The Dutch-born and -educated Kuiper was a precocious student who trained under astronomers such as Jan Oort, for whom the vast spherical Oort cloud of distant comets that envelops our solar system is named. Kuiper then worked at Lick Observatory in California, before planning to move to an observatory on the remote island of Java. He stopped at Yerkes on his way back eastward to spend a final year at Harvard University before sailing off. Kuiper's attitude impressed Struve, who was director of Yerkes at that time; Struve sent Kuiper a job offer that changed the course of astronomy in the Southwest from Texas to Arizona and beyond.

After amazing exploits by Kuiper in World War II that included important research in radar techniques and aiding the daring rescue of the great German physicist Max Planck, he became the second director of McDonald Observatory. Kuiper is best known today for his pioneering work in laying the groundwork for understanding both the structure and individual bodies at the outer edges of the solar system, now called Kuiper Belt objects. He went on to found the renowned Lunar and Planetary Laboratory at the University of Arizona in Tucson, which continues to be a world leader in solar system studies.

Struve, for whom the 82-inch telescope was later named, used it for his historic spectroscopic studies of stars and gas clouds in the Milky Way. A fourth-generation astronomer and veteran of the White Russian Army (the losing side in the Bolshevik Revolution), Struve was the first director of the combined observatory. By all accounts, he and his wife charmed the local populace, creating an environment in which local citizens took great pride in the presence of what must have seemed like a rather exotic group of scientists suddenly introduced into the town's daily life.

The next large telescope built at McDonald was the 107-inch (2.7-meter) Harlan J. Smith Telescope, named for the first astronomer from the University of Texas at Austin to direct the observatory, following the expiration of the original thirty-year agreement with the University of Chicago. Completed in 1968 with crucial funding support from NASA, the Smith Telescope was the third largest in the world when dedicated. It was the first large telescope to have its pointing system be automated successfully for computer control. The primary mirror in the Smith Telescope also survived a disturbing incident wherein it was shot by a deranged employee with a gun, as a result losing a tiny fraction of its light-gathering glass surface in the process—the stuff of west Texas legend!

The dominant telescope today at McDonald, both scientifically and visually on the summit of nearby Mount Fowlkes, is the 9.2-meter

(30-foot) Hobby-Eberly Telescope (HET), a bold effort to build a very large, specialized telescope for a relatively moderate cost ($13.5 million when complete). Major partners in the HET include Pennsylvania State University (home for several of its key early supporters), Stanford University, and two German universities.

Dedicated in 1997, the HET is a segmented-mirror telescope, similar to the two Keck telescopes built a few years earlier on Mauna Kea in Hawaii. Its primary mirror consists of 91 hexagonal segments that are each 1 meter (39 inches) in diameter, supported by a geometric ball-and-strut structure with more than 1,700 struts and 383 nodes, which, along with an active control system, ensures that the ensemble of mirrors forms a perfect reflecting paraboloid. This large, 80-ton structure turns on amazingly powerful air bearings. Some of the cost savings in the HET results from the fact that the HET itself does not move, but rather relies on the rotation of the Earth to move objects into its field of view. A downside is that this design both limits the effective aperture of the telescope to 9.2 meters at any one time, a loss of more than 35 percent of the maximum light-gathering power of the full 11-meter diameter of the entire mirror array, and restricts the amount of sky that the telescope can access.

The segmented-mirror approach, based partly on the technology pioneered by the two Keck telescopes, resulted in the need for an odd-looking alignment tower next to the telescope, where a laser system inside the 10-foot-diameter dome helps keep the mirrors precisely aligned with one another.

The creative design of the HET dome also helped save money: Rather than relying on dozens of steel workers and associated heavy machinery to slowly build it up, a crew of just three men used rivet guns to attach prefab aluminum sheets to a commercially purchased geodesic dome. The entire structure was then lifted in one piece by a 300-foot-tall crane and set in place in about ten minutes. The design of the HET served as the basis for the South African Large Telescope, and the approach of using segmented mirrors to make a large light-gathering surface is being applied in designs for future extremely large telescopes that will span 30 meters or more currently in development by teams in the United States, Canada, Japan, and Europe.

The primary mirror of the Hobby-Eberly Telescope (HET) at McDonald Observatory. The mirror is made up of 91 segments and has an effective aperture of 9.2 meters.

The HET is known primarily for its ability to produce very-high-resolution spectra of the light from both stars and distant galaxies, which makes it an excellent machine for searching for extrasolar planets and for measurements of dark energy, the subject that inspired an exciting new instrument for the telescope called HETDEX (a device that will enable measurement of the spectra of up to 10,000 galaxies during a single exposure).

Beyond the HET, the observatory also houses a laser-ranging station capable of measuring the Earth-Moon distance and the drift of Earth's continents to better than a centimeter, as well as an automated telescope that responds to newly sighted gamma-ray bursts, powerful explosions far beyond the Milky Way.

One of the newest facilities at McDonald is a compact 1.2-meter (48-inch) robotic telescope housed in an unusual-looking clamshell dome. Called MONET, for "MOnitoring NEtwork of Telescopes" (a verbal stretch!), this project is an international partnership between McDonald Observatory, the University of Göttingen in Germany, and the South African Astronomical Observatory; a second MONET telescope will be located northeast of Cape Town. This pair of telescopes will enable searches for extrasolar planets and studies of variable stars and active galactic nuclei (AGN) and will serve as an educational tool for students in the United States, Germany, and South Africa.

For the Public

Beyond its landmark large telescopes and rich scientific history, McDonald Observatory has been a leader in astronomy education and public outreach, including popular stargazing parties, a bimonthly magazine, and (most famously) *StarDate,* the longest-running national radio program on science.

The centerpiece of any visit is the modern Frank N. Bash Visitor Center (named for the widely admired director of the observatory from 1989 to 2003). This graceful, low-slung building houses a 300-seat amphitheater, an exhibit hall largely centered on the science of spectroscopy, a classroom, a 90-seat orientation theater, a gourmet-style cafe, and a large gift shop. Built from distinctive red sandstone mined in Pecos, Texas, the overall geometry of the $7 million project is based on a series of three spirals—the entrance plaza, the outdoor eating area and the amphitheater—with the building situated at the center. The spirals and circles reference ancient Native American ruins found in the Southwest.

The visitor center is open daily from 10:00 A.M. to 5:30 P.M. (additional hours on star-party nights), closed only on Thanksgiving, Christmas, and New Year's Day, with an admission fee required. Ninety-minute guided tours are offered every day at 11 A.M. and 2 P.M., along with live solar viewing via a video feed from a small telescope in an exterior building nearby, when allowed by clear skies. The HET is open for self-guided tours of its visitor galleries during daytime hours.

Twilight and evening stargazing programs are available at a moderate cost and at family rates for five or more. A combination pass can be purchased for a discount on a daytime/nighttime visit.

For a more elaborate treat, both the 82-inch and 107-inch telescopes are open on the order of once per month for private dinners and guided observing for $50 to $75 per person, making these telescopes

The Frank N. Bash Visitors Center opened in 2002. The interior houses exhibits, a theater, and a cafe. This photo shows the Sundial Court (front), the patio of the StarDate Cafe (left), the Rebecca Gale Telescope Park (right, rear), and the amphitheater (center, rear).

among the largest research telescopes in the world routinely available for public use.

A recorded message detailing hours, prices, and other information about visiting the observatory is available at (877) 984-7827. For other inquiries, send e-mail to info@mcdonaldobservatory.org.

The observatory is also one of the few to publish its own popular-science magazine. The bimonthly *StarDate* (with a circulation of about 8,000) began publishing in 1986, while the ubiquitous *StarDate* radio program runs on 350 stations nationwide, and its Spanish-language cousin, *Universo* (provided free of charge, see www.radiouniverso.org), runs on about 145 stations. Call 1-800-STARDATE for more information.

For Teachers and Students

McDonald Observatory staff conduct a variety of K–12 teacher workshops on-site, primarily in the months of June and July for four days/three nights and for shorter periods during the school year. The courses offer twenty hours or more of continuing-education credit, using inquiry-based activities aligned with state and national standards. Teachers are immersed in the environment, taking all their meals at the observatory, meeting visiting astronomers, and learning basic astronomy skills. Typical topics include the solar system, the age of the Milky Way, and light and optics. Some programs require teacher-paid fees; others are covered by foundations and grants. See www.mcdonaldobservatory.org for the latest details.

With the help of a rotating cast of college student interns, McDonald staff members consistently update a challenging public information project known as "What Are Astronomers Doing?" This interesting feature on the observatory's Web site provides both scientific synopses of projects and personality profiles of the researchers undertaking them—a challenging task by its very nature, but well executed.

A Talk with Anita Cochran

Assistant director, McDonald Observatory

Comet scientist Anita Cochran has served McDonald Observatory for three decades in a variety of positions with increasing responsibilities for the institution and its programs. She received her undergraduate degree from Cornell University and then moved to the University of Texas at Austin as a graduate student, where she met her future husband and fellow longtime McDonald Observatory planetary astronomer, Bill Cochran. She has spent her career studying the origins of our solar system.

What are your duties at McDonald Observatory?

I run the committee that judges proposals from astronomers for telescope observing time, and then I schedule our telescopes based on the rankings. I also run the computing services for the observatory and the University of Texas astronomy department, oversee various other technical projects, and work on reports and whatever other tasks are required by the director.

When did you first join the staff and in what sort of job? What are your earliest memories of the observatory?

Anita Cochran, assistant director, McDonald Observatory.

I came to the University of Texas as a graduate student in 1976. I received my PhD in the fall of 1982 and joined the research staff of McDonald Observatory in January 1983. I became assistant director in January 2003.

But my first visit to McDonald Observatory came in March 1977, when I was a graduate student. I was sent out to the observatory to learn to use the 36-inch telescope to do photometry of Uranus and Neptune to determine their rotation periods. Having grown up on the Atlantic Seaboard, my first impression was that the landscape surrounding the observatory was brown and barren. I have since grown to believe it is beautiful and wonderful.

Having come out to the observatory from Austin, I was not prepared for how cold it can be on a mountain at 7,000 feet above sea level in Texas. I shivered so hard that the person teaching me to use the telescope had to tell me to stop shaking the platform. Despite that, I fell in love with observing and with watching the night sky. In March, the Milky Way stretches overhead in the middle of the night and it is quite beautiful.

How have your duties changed over the years?

Having started as a grad student, I have spent my thirty-one years at McDonald moving up in responsibility. Once I had my doctoral degree, I first became a postdoctoral student, where all I had to do was research, and someone else worried about funding. As I moved up, I had to secure funds, supervise instruments being built, etcetera. Now, I have important responsibilities in the operations of the observatory and in budget planning and personnel supervision.

My career has taken turns I never would have imagined, but helping to lead the observatory is rewarding because it helps ensure that good science is always being done. It is also encouraging to me that half or more of our current graduate students are women; it was rare to find more than a few females in my classes as a grad student.

Why study comets?

We study comets in order to understand the conditions in our solar system at the time the planets formed. I have used my observations to fit together many pieces of the puzzle of the early solar system, including helping to figure out that comets basically come from two distinct chemical reservoirs in the solar system. The fact that we see that all comets are not the same tells us that different parts of the solar system experienced different conditions when comets were formed. Since comets have changed very little since then, they offer special insights into these conditions as opposed to those we can infer from observing the planets, which have changed a great deal since they were formed (as a result of both internal and external events).

What were the most interesting times at McDonald Observatory from a scientific standpoint?

My two most exciting experiences at a telescope both had to do with comet impacts. In 1994, I was part of the team at McDonald which watched the pieces of comet Shoemaker-Levy 9 impact Jupiter. I had never imagined I would see the image of Jupiter change before my eyes! It was fantastic. Then in 2005, I was at the Keck 1 telescope watching the impact of the *Deep Impact* spacecraft with comet 9P/Tempel 1 and taking spectra with the HIRES spectrograph.

In contrast, when we were observing Hale-Bopp in March 1997 at McDonald, the comet was rising in the east in the morning and coming up tail first. My observing partner, Ed Barker, and I would take turns guiding the telescope and running the instrument with which we were making measurements, while the other would grab the binoculars and go running out to the catwalk to look at the comet. Ironically, it looked much better in binoculars than in a 107-inch telescope!

What would you say is the most distinctive thing about McDonald Observatory?

Because McDonald Observatory is owned and run by the University of Texas at Austin, and its observers make frequent trips to the observatory, there is a great sense of family and personal history. I have grown up, astronomically speaking, with these people. That makes it very special to me.

McDonald Observatory has incredibly dark skies because it is a long way from anywhere. It is beautiful, mountainous, semi-arid country. People who work there are highly trained but love the small-town feel. We are a high-tech oasis in the middle of ranching country.

What knowledge or understanding do you hope that a visitor would take away from a trip to McDonald Observatory?

I would hope that a visitor would take away the fact that we are searching for knowledge, and that science is about posing questions and trying to figure out what kind of data we need to collect in order to answer those questions. I also hope that a visitor would find an interest in the sky and, even once they return to their homes, they would look up at the sky and ponder what they were looking at.

How do you see the observatory evolving over the next decade or two?

Unfortunately, I see our observatory in west Texas becoming less important for forefront science as we move to bigger and bigger telescopes. McDonald will still remain an important site for doing serious work and for training the next generations of astronomers. I hope that it will also remain an important destination for the public and continue to inspire people to appreciate the night sky.

Science Highlight

The Elements of Life

The first stars that appeared early in the life of the Universe, about 14 billion years ago, were built from the mix of chemical elements synthesized during the rapid expansion of matter following the Big Bang. This chemical soup consisted primarily of hydrogen and helium—the heavier elements (those with nuclei containing larger numbers of protons and neutrons) were absent at the beginning. How did these heavier elements (carbon, nitrogen, and oxygen), which are so basic to forming the building blocks of life, and other even rarer, heavy elements come to be made?

Understanding the origin of the complete set of elements comprising the periodic table represents one of the triumphs of modern physics and astrophysics: a combination of theoretical studies, laboratory measurements, and astronomical observations working in tandem. Over the past century, we have come to learn that the heavier chemical elements are synthesized in the interiors of stars via nuclear reactions that build the heavier, more complex elements from lighter elements.

One type of nuclear reaction involves "fusion" of lighter atoms to form heavier atoms in the hot, dense interior regions of stars. The interior of the Sun is powered by a series of nuclear reactions, which in the end fuse four hydrogen atoms into a single helium atom. This is the same set of reactions that on Earth can unleash the immense power of a hydrogen bomb. Element building can also take place via "bombardment" of atoms such as iron with fast-moving lighter atoms, such as helium. Production of heavy elements via bombardment has a terrestrial counterpart in particle accelerators used for research in physics, which can produce superheavy elements by impacting uranium nuclei with a stream of lighter particles accelerated to near the speed of light.

The cosmic nuclear reactions that produce heavy elements from lighter ones can take place at a stately pace, over hundreds of millions to billions of years at the centers of stars having masses within a factor of a few times our Sun. At the other extreme, heavy elements can also form in just a few seconds when a star ten to one hundred times more massive than the Sun explodes violently, becoming a supernova.

Supernovae produce heavy elements both by rapid fusion of lighter elements and by bombardment of newly formed heavy elements by fast-moving hydrogen, helium, and other light nuclei. Element production at intermediate speeds (over several million years) can also take place via bombardment in hot, nuclear-burning "shells" below the surfaces of red giant stars (stars of mass similar to the Sun whose surface temperatures are half that of the Sun, but whose radii are thirty to one hundred times larger).

Over time, chemical elements produced in stars via a variety of nuclear reactions are expelled into the gaseous medium between stars and galaxies, enriching the initially hydrogen- and helium-dominated gas with heavier atoms. The mechanisms for expulsion range from the explosive (via supernovae, which can eject atoms at tens of thousands of kilometers per second) to the comparatively gentle (by stellar winds blowing at a few hundred kilometers per second, which carry off heavy elements transported by convective currents from stellar interiors to their outer atmospheres). Next-generation stars are formed from this heavy-element-enriched gas and, in turn, build more heavy elements, eventually ejecting these newly formed heavy elements into the interstellar or intergalactic medium. All of the chemical elements on Earth are by-products of 7 billion to 10 billion years of this grand set of processes—in supernovae, the cores of stars a few times the mass of the Sun, and in the shells of red giant stars. In a very real sense, humans and other life forms owe their existence to the buildup of chemical elements in stars, their eventual ejection into the vast gaseous medium between the stars, and their incorporation into new stars and their associated planetary systems.

Astronomers at the McDonald Observatory have played key roles in assembling the story of how chemical elements are made, in what quantity, when, and in what kinds of stars, how they are transported to stellar surfaces, and how they eventually make their way into the interstellar medium to form new generations of stars. Their main tool for exploring how the mix of chemical elements evolved over the history of our Milky Way galaxy—and indeed over the history of the Universe—is the spectrograph.

This instrument makes use of an optical element (such as a prism or grating) to disperse the light from distant stars into its component colors. A simple, small prism held up to the Sun will produce a "rainbow" of colors. A large prism or a grating, such as those comprising the basic element of a spectrograph, will break light up into finer colors—fine enough to reveal not only broad swaths of colors, but a diagnostic pattern of extremely sharp, dark lines superimposed on a smooth colored background. These spectral lines represent unique "fingerprints" produced as collections of atoms from a given element absorb photons of very particular wavelengths or colors.

By mounting a spectrograph at the focal plane of a large telescope, such as McDonald Observatory's 2.7-meter Smith Telescope or the 9.2-meter Hobby-Eberly Telescope, it is possible to record the complex pattern of spectral lines arising from the mix of chemical elements present in the relatively cool outer atmospheres of stars. The role of the telescope is basically that of a light bucket, designed to collect light of all wavelengths from often distant, faint stars; the role of the spectrograph is to break up the light collected by the telescope into finely divided colors or wavelengths.

By measuring the wavelengths of individual spectral lines, it is possible to identify which elements are present in the atmosphere of a target star by comparison of the patterns observed in a stellar spectrum with the pattern of spectral lines cataloged by measurements of heated samples of elements in a laboratory setting. By measuring the strengths of individual lines arising from each element, it is possible to determine their abundances. The latter step requires not only measurement of line strengths, but modeling of the temperature and pressures in the atmospheres of stars. This complex process, which depends on computer simulations, finds ultimate verification in its ability to successfully match the observed pattern of line strengths for the lines of all elements.

Scientists at McDonald have carried out spectroscopic measurements of stars spanning a wide range of masses, ages, and evolutionary states. From these measurements, they are able to trace how the abundances of different elements evolve over time by comparing the abundance patterns in stars formed early in the history of the Milky Way, when heavy elements were relatively rare, with those of more recently formed stars whose chemical abundances reflect the contributions to the current "mix" resulting from the ejecta from multiple generations of stars.

By examining the chemical patterns in stars of a given mass, but captured in different evolutionary states, McDonald astronomers have begun to develop an empirical understanding of what kinds of elements are produced by stars of what mass and at what evolutionary state.

From their measurements and those of other researchers, we now have a fairly good idea of how and in what quantity elements are produced in stellar cores and shells; how and at what rate these elements are "mixed" from interior to shell, from shell to atmosphere and vice versa; and how newly formed elements are ejected into the interstellar medium. These same studies provide powerful insights not only into how elements are made, but into the structure (temperature and pressure) of the hidden interior regions of stars—crucial to improving our understanding of the basic physical processes that govern the stars' evolution and eventual death.

Exterior view of the Vatican Observatory Telescope at night with shutter open.

Mount Graham International Observatory

Safford, Arizona

The Most Powerful Ground-Based Telescope in the World

The most isolated major observatory in Arizona can be found in the southeastern part of the state on rugged Mount Graham near Safford, where an eclectic trio of telescopes sits amid a heavily wooded wilderness at the end of one of the more imposing observatory access roads in North America.

Mount Graham International Observatory began life in 1984 with an act of Congress that designated 3,500 acres for an astronomical observatory and another 62,000 acres as a wilderness study area. Located at a steep elevation of 10,720 feet, Mount Graham reaches into heights where the light-blocking effects of water vapor in Earth's atmosphere become much less of a problem, because there isn't much vapor above the telescope, at least on most days.

Web site
www.eac.edu/discoverypark/mgio.shtm

Phone
(928) 428-6260

E-mail
discoverypark@eac.edu

Address
Visitor access via:
Eastern Arizona College Discovery Park Campus
1651 West Discovery Park Boulevard
Safford, AZ 85546

However, the choice of this location triggered a classic environmental and social clash between the University of Arizona and its scientific partners versus the San Carlos Apache Tribe (advocating their sacred claims to the mountain) and a coalition of environmentalists concerned that the construction of the observatory amid the spruce-fir forest would lead to the demise of the Mount Graham red squirrel. This endangered member of the rodent family is known for its fluffy tail and lack of vocalization—characteristics likely due to their evolutionary isolation from their low-altitude brethren.

The opponents filed some forty lawsuits, eight of which reached a federal appeals court, before another act of Congress pushed the project forward. (The red squirrel population survived the construction of the telescope, as well as major fires in 1996 and 2004, and seems to be hovering near its early 1990s population of 200–250.)

After decades of anticipation, the mammoth Large Binocular Telescope (LBT), the undisputed centerpiece of Mount Graham, is ready to assume its place as the largest operational telescope in observatory-rich Arizona. When the combined size of both of its 8.4-meter (27.6-foot) mirrors on a single mount is factored into the equation, the LBT is the world's largest ground-based telescope, equivalent in light-gathering ability to a telescope with a single 11.8-meter (38.7-foot) primary mirror, a bit larger than other contenders in the Canary Islands and on top of Mauna Kea in Hawaii.

The LBT has a moving weight of 600 tons, hung from two 46-foot-wide, C-shaped main bearings. This red-painted behemoth is surrounded by a 2,000-ton rectangular enclosure that is sixteen stories tall, its boxy exterior marked by several odd-looking air exhaust and ingestion tubes.

The LBT was conceived as a pathfinder in several ways, from its double set of mirrors to a suite of instruments permanently mounted at a variety of different focal points. These instruments are available to be switched into the beam of starlight much more quickly than on most other telescopes.

Beyond the light-gathering sensitivity of its two mirrors operating in a parallel observing mode, the fact that the two LBT mirrors are separated from center to center by a distance of 47 feet gives astronomers

Exterior of the Large Binocular Telescope with shutters open.

the ability to coherently combine the incoming light from the two giant "light buckets" with great precision—an observing mode known as interferometry (see pages 78–79). Used primarily to resolve fine details in the structures of planetary disks and galaxies, and to separate individual stellar point sources in densely crowded clusters, this technique enables the LBT to achieve a resolution equivalent to a telescope with a 23-meter (75.5-foot) primary mirror—nearly as large as the largest future ground-based telescopes being imagined today.

Among many possibilities, this "binocular" capability may enable the LBT to take actual images of extrasolar planets, which until now

Interior of the Large Binocular Telescope.

have been observed by the secondary effects they cause on their parent stars (using their gravity to tug the star back and forth slightly during their orbits or blocking the star's light by a tiny fraction when passing in front of them), and to resolve the disks of gas and dust from which planets form early in the lifetimes of their parent stars.

The LBT is an international partnership between the University of Arizona's Steward Observatory and the countries of Italy and Germany (25 percent each), along with Ohio State University and a number of other universities that each purchased a modest share of LBT. The project, which began essentially in 1989, has cost a total of $150 million. A less-than-ideal funding profile over the years and a major political and environmental fight over the site during the 1990s stretched the project out, painfully at times.

The mirrors are the product of an international collaboration as well, made from Florida sand that was turned into 40 tons of glass in Japan and then formed into polished 18-ton mirrors in the University of Arizona mirror lab. The mirrors made the tortuous three-day truck trip up the mountain at 1 mph in 2002 and 2005, respectively. One unique feature of the mirrors (and their mounts) is that they are designed to have their reflective surfaces recoated with a thin aluminum film (to restore high reflectivity) while they are in place, rather than being moved to a separate chamber in an inherently dicey maneuver that is one of the least-welcomed maintenance tasks for any large telescope.

The LBT is also unique in that it has taken on the challenge of making its large (0.9-meter or 36-inch) secondary mirrors flexible so they can compensate for distortions in image quality wrought by turbulence in Earth's atmosphere; it does this by using a technique known as adaptive optics (see page 10). Because of the Gregorian design of the LBT, these complicated mirrors are concave (curved inward) rather than the much more common convex design. Only 1.6 millimeters (0.06 inch) thick, each mirror is backed by 672 small piezoelectric magnets that can push or pull on the surface one thousand times per second when subjected to electrical signals emanating from a device known as a wavefront sensor, which monitors in near real time the ongoing changes in image distortion caused by the atmosphere. The testing process for the mirrors included a historic tie back to George Ellery Hale, the famous founder of Mount Wilson and Palomar observatories: A venerable solar-observing tower at Arectri Observatory near Florence, Italy, built by Hale, was modified and used to test the LBT secondary mirrors.

Each large primary mirror of the LBT has a large 36-megapixel digital camera at its prime focus—one optimized for blue-tinted starlight and one optimized for red. The mirror with the blue camera above it saw first light in October 2005 and started its first major science observations in January 2007; the red camera opened for science in November 2007.

The first binocular observations by LBT were achieved in early 2008. Keeping the two cameras tightly aligned with each other is an amazing feat: Their positions must be controlled to a precision of 0.1 millimeter (0.004 inch) in order for light from the two telescopes to be combined coherently, a requirement for creating high-resolution, interferometric images. This must be done in the face of the variable "sagging" of the overall telescope structure (due to Earth's gravity) by as much as 20 times that amount, as the massive telescope structure rotates from the zenith toward the horizon.

Early science results from LBT include detailed maps of the star types in nearby small galaxies and studies of a tiny dwarf galaxy with a very odd elongated shape, implying either that it is gravitationally distorted as a result of a trip through our Milky Way or that it suffered a very unusual birth early in the history of the Universe.

But it is studies of distant planets around other stars where the LBT is likely to make some of its major marks. A special planet-hunting instrument fed by optical fibers will sit in a temperature-controlled trailer at the base of the telescope. It aims to make such precise measurements of the gravitational tugs of planets on their parent stars that the support trailer is outfitted with supersensitive accelerometers capable of measuring tiny conflicting signals such as the force of ocean waves hitting the distant coast of California, more than 500 miles away.

Mount Graham is also home to two other somewhat unusual astronomy facilities. The Vatican is not the first place that many people would associate with ongoing scientific study of astronomy. However, the Vatican has a long history of supporting (and fighting with) astronomy stretching back four hundred years, to the first use of the telescope for astronomy by Galileo Galilei.

The Vatican has supported astronomical research at sites ranging from Rome to the pope's summer residence south of the city in Castel Gandolfo—and to the arid peak of Mount Graham in Arizona's Pinaleño Mountains.

Dedicated in September 1993, the Vatican Advanced Technology Telescope (VATT) was the first optical-infrared telescope on the mountain. The 1.8-meter (71-inch) primary mirror for the VATT was an early landmark accomplishment by the University of Arizona mirror laboratory, in its first attempt at spin-casting a mirror over ceramic molds inside a rotating furnace.

Asked often about the apparent conflict between religion and the quest for scientific knowledge, VATT directors characterize their purpose as fostering a "respectful dialogue" about the respective roles of the two in our lives. As such, the research carried out with the VATT is similar to the high-quality work with modest-sized telescopes around the world, ranging from studies of galaxy evolution to the properties of asteroids.

Exterior of the LBT.

The other major facility on Mount Graham also has some international roots. The Arizona Radio Observatory SubMillimeter Telescope (SMT) is a University of Arizona facility run by its Steward Observatory staff. But it originated as a cooperative project between Steward Observatory and the Max-Planck-Institut für Radioastronomie in Bonn, Germany, and it began life as the Heinrich Hertz Telescope. Designed to probe the same type of energetic radiation that most people know from their kitchen microwaves, the SMT is equipped with some instruments that are similar to those found on optical-infrared telescopes, while others are more analogous to those found on radio telescopes. It has a large 10-meter-wide (33-foot-wide) dish like a radio telescope but is housed in an enclosure more similar to those sheltering optical telescopes. The careful design of the composite-material structure that supports the main dish makes it the most accurate radio telescope ever built, according to the project staff; it has a 12-meter (39-foot) radio observatory "cousin" operated by the University of Arizona on Kitt Peak.

The microwave part of the electromagnetic spectrum is ideal for observing features as diverse as the chemistry of interstellar clouds that could give birth to stars to the remnants of radiation at the edge of the observable Universe, which marks the beginning of the cosmos as we know it, just 400,000 years after the Big Bang.

The telescopes on Mount Graham have survived both a period of mostly unwelcome notoriety and intense forest fires to reach a point where they are poised to deliver unexpected discoveries of all types.

For the Public

The official tour agent for Mount Graham International Observatory is a mixed-use science center, nature preserve, and conference facility known as Discovery Park, now operated by Eastern Arizona College. It is located in the small town of Safford, at the intersection of Routes 70 and 191 in southeastern Arizona. Most visitors will find themselves accessing 191 from Interstate 10, a 33-mile drive to the south.

Public tours of Mount Graham Observatory are conducted by advance reservation, beginning about mid-May through mid-November, weather permitting. Be aware—the 40-mile trip up a hilly and curving road to the summit and back is not for the fainthearted, and it requires a full-day commitment.

The observatory tours begin with a 9:00 A.M. arrival time at Discovery Park and an orientation briefing on the history and ecology of the mountain. The tour van leaves Discovery Park for Mount Graham Observatory at 9:30 A.M. and returns at approximately 4:30 P.M. The $40-per-person ticket price includes a sack lunch at the U.S. Forest Service's Columbine Visitor Center. Among the supplies recommended at the Discovery Park Web site (www.eac.edu/discoverypark/mgio.shtm) are sunscreen, water, aspirin, and (if needed) Dramamine.

Tours require a minimum of six people, and advance notice of two weeks or more is preferred. Because of the high altitude (10,720 feet), the tour is not recommended for people with heart conditions or respiratory problems, and children must be at least eight years old. To make reservations or for further information, call (928) 428-6260 or e-mail discoverypark@eac.edu.

Discovery Park in Safford also is home to a 20-inch (0.5-meter) telescope in its Gov Aker Observatory, a virtual-reality space shuttle ride through the solar system, a replica of an 1860s steam locomotive, a variety of nature trails, and one of the world's largest camera obscuras. The Discovery Park campus is free to the public Monday through Friday, 9:00 A.M. to 5:00 P.M., and Saturday evenings from 4:00 P.M. to 10:00 P.M.

For Teachers and Students

No-cost tours and guided science field trips are available to schools in the surrounding area. Boy Scout troops and Cub Scout packs are also encouraged to visit for special tours and for meetings using the surrounding nature habitat.

And More: Mount Lemmon SkyCenter

In addition to Mount Graham and several telescopes on Kitt Peak, the University of Arizona operates a significant observatory on Mount Lemmon, about 45 miles northeast of its Tucson campus in the Santa Catalina Mountains. The observatory was a radar base of the U.S. Air Defense Command until 1970, and it now hosts two 1.5-meter (60-inch) telescopes, plus a 1-meter (40-inch) telescope operated by the Korean Astronomy Observatory, as well as several smaller telescopes, at a relatively high altitude of 9,157 feet.

The telescopes on Mount Lemmon were pioneers in the field of infrared astronomy and—in partnership with a 1.5-meter (60-inch) telescope on nearby Mount Bigelow—continue a landmark search for nearby asteroids known as the Catalina Sky Survey.

Known in recent years for intensive astronomy camps for students, advanced amateurs, and Scouts, the facilities on Mount Lemmon are now part of a larger university effort called the Mount Lemmon SkyCenter, which includes "Discovery Days," summer camps, and other programs linked both to astronomy and to environmental themes, such as the university's world-class tree-ring science department and the greenhouse-like Biosphere 2 facility north of Tucson.

Since May 2008, the astronomy operations include a new nighttime public observing program on Mount Lemmon using a loaned 0.6-meter (24-inch) telescope; the SkyCenter hopes to replace this with a 0.8-meter (32-inch) or larger, which promises to be among the largest in the world available for regular public viewing. The cost is $48 per person for the evening program, with higher fees for an overnight experience. Call (520) 626-8122 for more information or go to skycenter.arizona.edu.

Acknowledgments

Thank you to all the interview subjects for your valuable time and unique perspectives; your input also helped shape the chapter content. Special thanks to all the hardworking current—and former—public information officers and public outreach staff of the observatories who contributed fact sheets, reports, press releases, photo ideas, historical tidbits, and important reviews of draft chapters: James Cornell, Dan Brocious, Dave Dooling, Kurt Anderson, Scott Kardel, David Finley, Rebecca Johnson, Frank Cianciolo, Steele Wotkyns, and Elizabeth Alvarez del Castillo, as well as Richard Green, director of the Large Binocular Telescope Observatory.

Thanks also to former University of Arizona Space Grant Intern Shiva Kiani, whose early work on material related to the fiftieth anniversary of Kitt Peak National Observatory contributed to the formation of the proposal for this book, and to our patient and constructive editor at the University of Arizona Press, Patti Hartmann.

Any remaining errors are the responsibility of the authors alone.

This book would not have been possible without the support and inspiration provided by Doug's wife, Paula, and their children, Zeke and Vega—may they continue to reach for the stars.

For Further Reading

General

Brunier, Serge, and Anne-Marie Lagrange. *Great Observatories of the World*. Ontario, Canada: Firefly Books, 2005.

Krisciunas, Kevin. *Astronomical Centers of the World*. Cambridge, U.K.: Cambridge University Press, 1988.

Fred Lawrence Whipple Observatory

Hogan, Donald. *Event History of Fred Lawrence Whipple Observatory*. Online publication of the Whipple Observatory: http://linmax.sao.arizona.edu/help/FLWO/flwohis.html

Kitt Peak National Observatory

Kloeppel, James E. *Realm of the Long Eyes: A Brief History of Kitt Peak*. San Diego: Univelt, 1983.

Sage, Leslie, and Gail Aschenbrenner. *A Visitor's Guide to the Kitt Peak Observatories*. Cambridge, U.K.: Cambridge University Press, 2004.

Lowell Observatory

Putnam, William Lowell, and others. *The Explorers of Mars Hill*. Canaan, N.H.: Phoenix Publishing, 1994.

Strauss, David. *Percival Lowell: The Culture and Science of a Boston Brahmin*. Cambridge, Mass.: Harvard University Press, 2001.

McDonald Observatory

Sebring, Thomas, and Joel Warren Barna. *West Texas Time Machine: Creating the Hobby-Eberly Telescope*. Fort Davis, Tex.: Little Hands of Concrete Productions for McDonald Observatory, 1998.

Winget, Karen Stewart. *Dear Visitor: Voices of McDonald Observatory*. Round Rock, Tex.: Caddem Publishers, 2003.

The Observatories of Sacramento Peak

Ramsey, Joan. *In the Beginning: Sacramento Peak, Sunspot, New Mexico*. Kearney, Nebr.: Morris Publishing, 1997.

Palomar Observatory

Florence, Ronald. *The Perfect Machine: Building the Palomar Telescope*. New York: HarperCollins, 1994.

Osterbrock, Donald. *Pauper & Prince: Ritchey, Hale, & Big American Telescopes*. Tucson: University of Arizona Press, 1993.

Sandage, Allan. "The First 50 Years at Palomar, 1949–1999: The Early Years of Stellar Evolution, Cosmology, and High-Energy Astrophysics." Reprint online from *Annual Review of Astronomy and Astrophysics*, 37 (1999): 445–486. http://nedwww.ipac.caltech.edu/level5/Sept03/Sandage/Sandage_contents.html

Woodbury, David O. *The Glass Giant of Palomar*. New York: Dodd, Mead, 1970.

Illustration Credits

Fred Lawrence Whipple Observatory

Converted MMT: Photograph by Howard Lester, MMT Observatory.

Observatory Ridge: Courtesy of Whipple Observatory.

VERITAS: Courtesy of Steve Criswell, Whipple Observatory.

Whipple scientists at public outreach event: Courtesy of Whipple Observatory.

Fred Chaffee: Courtesy of Robert T. Rood/University of Virginia.

Kitt Peak National Observatory

All photographs: Courtesy of NOAO/AURA/NSF.

Lowell Observatory

All six observatory photographs: Courtesy of Tom Alexander/Lowell Observatory.

Robert Millis: Courtesy of Lowell Observatory.

McDonald Observatory

All four observatory photos: Courtesy of Marty Harris/McDonald Observatory.

Anita Cochran: Courtesy of Anita Cochran/McDonald Observatory.

Mount Graham International Observatory

Exterior of Vatican Observatory Telescope at night with shutter open: Courtesy of David Harvey.

Exterior of Large Binocular Telescope with shutters open: Courtesy of John Hill and the Large Binocular Telescope Observatory.

Interior of Large Binocular Telescope: Courtesy of the Large Binocular Telescope Observatory; photograph by Marc-Andre Besel and Wiphu Rujopakarn.

Exterior of Large Binocular Telescope: Courtesy of David Harvey and the Large Binocular Telescope Observatory.

National Radio Astronomy Observatory Very Large Array

VLA in 2007: Photograph by John Littell; courtesy of NRAO/AUI.

VLA and transfer railroad tracks: Photograph by Matthew L. Abbondanzio; courtesy of NRAO/AUI.

VLA in snow in 2004: Courtesy of NRAO/AUI and Laure Wilson Neish.

VLA and visitor center: Courtesy of NRAO/AUI.

Rick Perley: Courtesy of Rick Perley/NRAO.

The Observatories of Sacramento Peak

Aerial view of Sacramento Peak Observatory: Courtesy of NSO/AURA/NSF.

Telescope spar in Evans Solar Facility: Courtesy of NSO/AURA/NSF.

Sunspot image: Courtesy of Friedrich Woeger, KIS, and Chris Berst and Mark Komsa, NSO/AURA/NSF.

Sloan Sky Survey 2.5-meter telescope: Courtesy of Fermilab Visual Media Services.

Sloan telescope enclosure building rolled over telescope: Courtesy of Fermilab Visual Media Services.

Sunspot Visitor Center: Courtesy of NSO/AURA/NSF.

Steve Keil: Courtesy of Steve Keil, NSO/AURA/NSF.

Palomar Observatory

All four observatory photographs: Courtesy of S. Kardel and Palomar Observatory.

Chuck Steidel: Courtesy of Chuck Steidel/Caltech.

Index

Note: Page numbers in italics refer to illustrations.

About the Authors

Stephen Strom is an astronomer emeritus at the National Optical Astronomy Observatory (NOAO). Until 2007, he served as associate director of the New Initiatives Office of the Association of Universities for Research in Astronomy (AURA) and in that position was responsible for overseeing NOAO's role in designing a 30-meter (98-foot) optical-infrared telescope (TMT) currently being carried out by a partnership involving the University of California, Caltech, and the Canadian Astronomical Community. The author of more than 200 research papers, his interests focus on the formation of stars and planetary systems. Strom has also held appointments at Harvard University, the State University of New York at Stony Brook, and the University of Massachusetts at Amherst (where he served as chair of the five-college Astronomy Department for nearly fourteen years). He has provided photographs for three collaborative works published by the University of Arizona Press: *Secrets from the Center of the World*, with Joy Harjo; *Sonoita Plain: Views from a Southwestern Grassland,* with Carl and Jane Bock; and *Tséyi'/Deep in the Rock: Reflections on Canyon de Chelly*, with Laura Tohe. He also provided photographs for the recently published *Otero Mesa: Preserving America's Wildest Grassland* (University of New Mexico Press), with Greg McNamee and Stephen Capra.

Douglas Isbell is the U.S. national contact for the International Year of Astronomy 2009, and an astronomy- and space-science-related communications professional with more than 20 years of experience at the National Optical Astronomy Observatory in Tucson, NASA, and the commercial space industry. Isbell has a proven ability to communicate highly technical information to a wide audience in a compelling manner. He is an award-winning writer and has been a frequent guest and spokesperson on television and radio. During his career, Isbell has planned and executed comprehensive media campaigns for topics of major worldwide interest, such as the Mars Pathfinder landing and the launch of the Cassini mission to Saturn.